CLINICAL ANATOMY MADE RIDICULOUSLY SIMPLE

Stephen Goldberg, M.D.
Associate Professor
Department of Cell Biology and Anatomy
University of Miami School of Medicine
Miami, Florida 33101

MedMaster, Inc., Miami

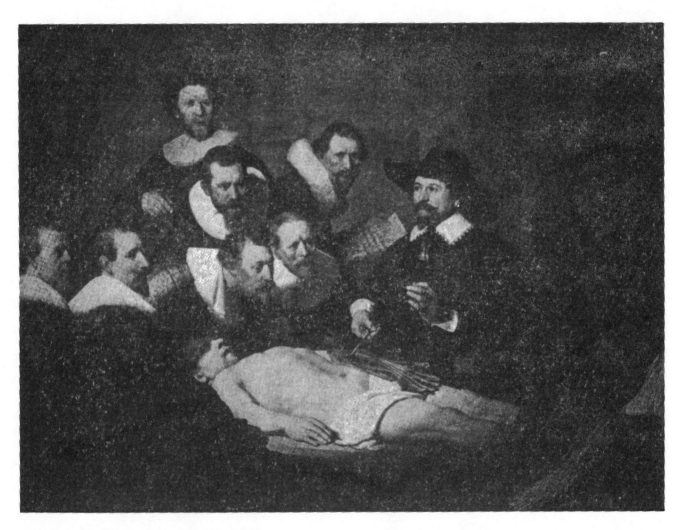

Copyright © 1984, 1991 by MedMaster, Inc.
Second printing, 1984; Third printing, 1986; Fourth printing, 1987; Fifth printing, 1987; Sixth printing, 1988; Seventh printing, 1989; Eighth printing, 1990; Ninth printing, 1991; Tenth printing, 1992; Eleventh printing, 1992; Twelfth printing, 1993; Thirteenth printing, 1994.

ISBN #0-940780-02-X

Made in the United States of America

Published by MedMaster, Inc.
P.O. Box 640028
Miami, FL 33164

"The Anatomy Lesson", Rembrandt (above), modified on the front cover by Sixten Netzler.
Anatomical figures by Albinus on inside of front and back covers.

TO DOCTORS HUGH BILLER, MARTIN BRODY, AND ALLEN ROTHMAN
IN GRATITUDE FOR THEIR SURGICAL AND INTERPERSONAL SKILLS

PREFACE

The purpose of CLINICAL ANATOMY MADE RIDICULOUSLY SIMPLE is to teach medical students and other medical professionals in a special way that enables the student to rapidly learn, retain and review anatomy.

It is difficult for the student to grasp and integrate the overwhelming detail inherent in this important subject, as a result of the reduced time allotted to the study of anatomy. This problem is especially acute where anatomy is taught according to body regions and where large reference texts are used alone. The student often remembers isolated points for exam purposes without adequately understanding the overall picture. This book is not intended to replace standard reference texts but rather to be read as a companion text. Supplementing the standard texts with a book of this nature will enable the student to learn anatomy more rapidly with better retention.

The key features of my book are:

1. I employ memory aids, particularly those involving humor and ridiculous associations, to immediately implant in mind the anatomy under study. Such devices have long been known to be effective but have seldom been employed in the medical literature, perhaps because they have been felt to be "unprofessional". Such methods work, however, and should be considered for broader use in medical education. My admittedly sophomoric attempts at humor in this book do not imply any disrespect for the field. They are a serious gesture aimed at delivering important information in an effective manner. I hope that readers will find such amateurish attempts either amusing, embarrassing, or insulting to the intelligence, as any of these reactions will more firmly affix the material in mind.

2. Memory aids are applied mainly to teach **anatomical structures** rather than their names. Names will come naturally if the structures are known. My approach is to purposefully distort the actual anatomy into ridiculous scenes or schematic views that can be easily grasped. From there, the student can make a smooth transition to the actual anatomy. I have largely avoided the type of mnemonic where the first letters of the names in a long list are converted into a clever ditty. Such lists teach only names, and become confusing when a student attempts to remember too many.

3. The book is predominantly organized by systems, as it is easier to learn anatomy by systems than by regions. There is a separate regional table of contents, however, for students whose courses are organized by regions, which is the case in most medical schools. Also, the final chapter covers key regional points.

4. The text is clinically oriented.

5. The figure legends and text have been fused into one continuum, to avoid unnecessary duplications and distracting jumps from text to figure legend and back again. The words in the index are assigned according to figure number rather than page number. For example, "Brachial vein, 6-39" means that the term "brachial vein" is found in illustration 6-39 and/or in the text between sections 6-39 and 6-40.

6. A small glossary is included for terms not defined in the text.

This brief book obviously cannot cover every anatomical point with clinical relevance to the radiologic or surgical subspecialties. Structures or structural relationships that have little **functional** importance may have **clinical** importance if only for their use as landmarks. This book is not intended for the subspecialist. It tries to give a broad initial perspective of the most important anatomical points. Subtle points that relate to the subspecialist will be obtained from more detailed texts, perhaps at the time of subspecialty training. The book also provides a rapid review for medical Boards and other exams which emphasize clinically relevant aspects of anatomy.

I welcome suggestions from the many excellent teachers of anatomy, as well as students, who have successfully used similar techniques in their programs. Helpful suggestions will be included in future printings (if the book is fortunate enough to reach that point), with due credit to the contributors.

I thank Drs. Donald Cahill, Humberto Valdes, and Phillip Waggoner for their helpful comments and Beryn Frank for editing the manuscript. Text illustrations are mainly by the author, with the assistance of illustrations from Diane Abeloff's "Medical Art Graphics for Use", Williams and Wilkins, 1982. The cover illustration is by Sixten Netzler. I thank Dr. Ming X. Wang of Harvard Medical School for the "Mid-ear Strong Man" analogy (fig. 16-1) and for the "Soccer Player" analogy (fig. 17-10), as well as Dr. Robert Davies for several other memory aids.

Stephen Goldberg

CONTENTS (SYSTEMIC)

CONTENTS (REGIONAL)

CONTENTS

CONTENTS

CHAPTER 1. ORIENTATION

Fig. 1-1. Sectional cuts.

Fig. 1-2. Axes, as seen with the subject in the anatomical position. For the neck and trunk, the terms dorsal and ventral are synonymous with posterior and anterior, respectively.

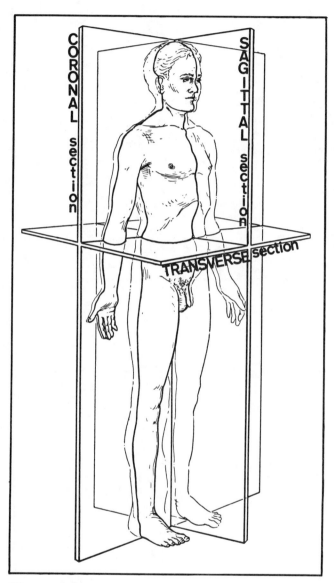

Figure 1-1

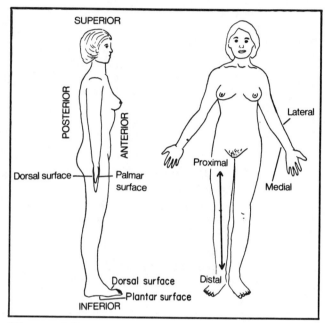

Figure 1-2

Fig. 1-3. Movements of upper and lower limbs.

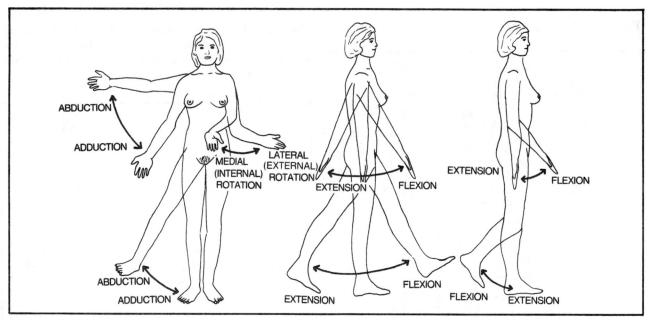

Figure 1-3

Fig. 1-4. Movements of hand and foot. In ankle eversion, the sole of the foot moves out; in inversion the sole moves in.

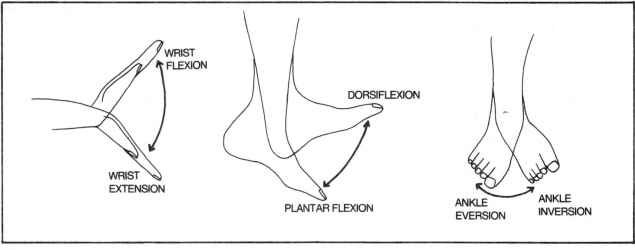

Figure 1-4

Fig. 1-5. Movements of fingers and thumb. Flexion-extension.

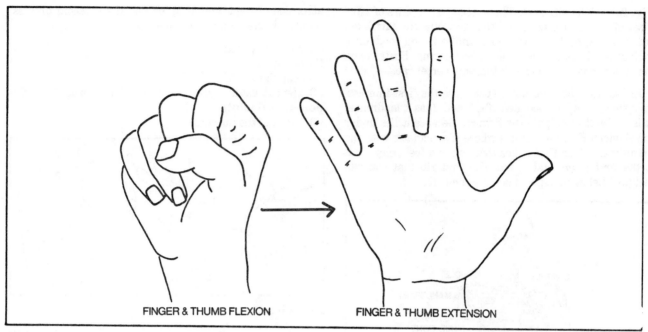

FINGER & THUMB FLEXION   FINGER & THUMB EXTENSION

Figure 1-5

Fig. 1-6. Movements of fingers and thumb. Abduction-adduction and opposition. In finger abduction, the index, ring, and small fingers move away from the middle finger.

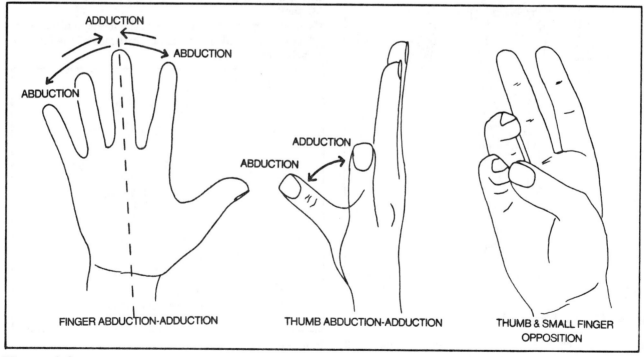

ADDUCTION

ABDUCTION

ABDUCTION

ADDUCTION

ABDUCTION

FINGER ABDUCTION-ADDUCTION   THUMB ABDUCTION-ADDUCTION   THUMB & SMALL FINGER OPPOSITION

Figure 1-6

CHAPTER 2.   THE SKELETAL SYSTEM

Fig. 2-1.   The rib cage. There are twelve pairs of ribs. Most of them fuse with the sternum (via rib cartilage) but ribs 8,9, and 10 don't quite make it and grab onto each other. Ribs 11 and 12 are free floating. Posteriorly, the ribs connect with the 12 thoracic vertebrae.

In locating the intercostal spaces, use the first intercostal space (the space between ribs 1 and 2) as a landmark. It is difficult to palpate the first rib because it lies under the clavicle. Feel the clavicle where it articulates with the manubrium. The first space that you can feel below that is the first intercostal space. The first rib that you can feel (just below this space) is the second rib.

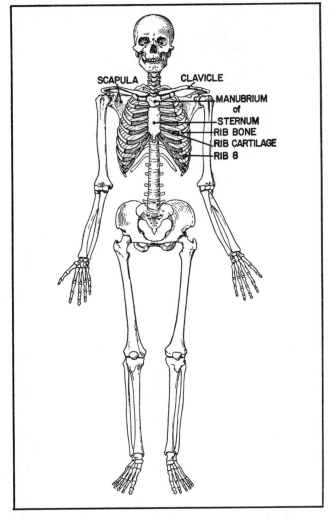

Figure 2-1

Bones Of the Upper Extremity

Fig. 2-2.   A right scapula. The scapula resembles a bull's head with the horns on backward. Its horns are the

acromion of the scapular spine and the coracoid process. Its neck is the neck of the humerus.

(1) coracoid process
(2) acromion
(3) scapular spine
(4) glenoid cavity - for insertion of humerus (the **glenohumeral joint**)
(5) neck of the humerus

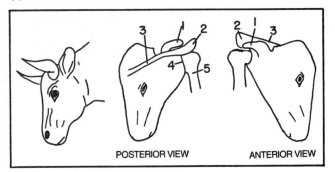

Figure 2-2

Fig. 2-3.   A right humerus.

(1) head of humerus
(2) anatomical neck
(3) surgical neck - fractures may occur through the anatomical or surgical necks but are particularly common through the surgical neck, especially in the elderly
(4) shaft
(5) medial epicondyle - **flexors** of the forearm attach here
(6) lateral epicondyle - **extensor muscles** of the forearm attach here

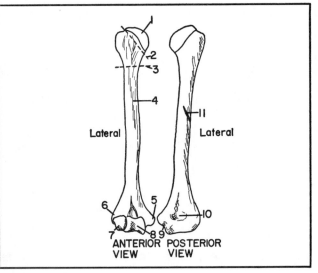

Figure 2-3

(7) capitulum - caput, or head - articulates with the head of the radius

(8) trochlea - dumbell-shaped, articulates with the ulna

(9) ulnar nerve sulcus - crossed by the ulnar nerve; responsible for the horrendous feeling obtained on striking this area (hitting one's "funny bone")

(10) olecranon fossa - the olecranon of the ulna fits in here

(11) radial sulcus - a groove containing the radial nerve. Fractures of the humerus may thus injure the radial nerve.

Fig. 2-4. Bones of the forearm and hand. The ulna resembles a monkey wrench. The fingers contain digital bones (**phalanges**) while the rest of the hand contains 5 **metacarpal** and 8 **carpal** bones. The carpal bones are arranged in two rows. Starting at the radius (the male lover), the carpal bones project the mnemonic:

(1) SCARED (scaphoid)
(2) LOVERS (lunate)
(3) TRY (triquetrum)
(4) POSITIONS (pisiform)
(5) THAT (trapezium)
(6) THEY (trapezoid)
(7) CANNOT (capitate)
(8) "HAND"LE (hamate)

"The trapezium supports the thumb. The trapezoid's on the inzoid."

How do you remember that the trapezium comes before the trapezoid? Ans. They are in alphabetical order.

The nose of the male lover (radial tuberosity) is the site of insertion of the powerful biceps tendon.

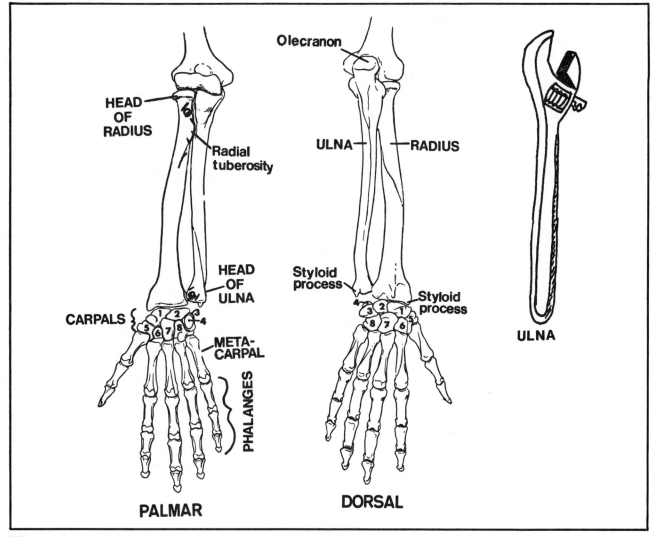

Figure 2-4

Fig. 2-5. Colle's fracture. Dinner fork deformity of the wrist following a fall on an outstretched hand.

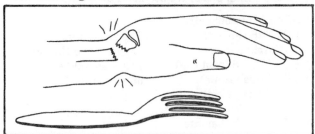

Figure 2-5

Vertebrae

The vertebral column usually contains 7 cervical vertebrae, 12 thoracic vertebrae (for 12 thoracic ribs), 5 lumbar vertebrae, a sacrum, and a coccyx.

Fig. 2-6. The vertebra and 31 pairs of spinal nerves. All the spinal nerves run under their respective vertebrae, except for the cervical nerves which run above their respective vertebrae. This leaves a gap between C7 and T1 vertebrae, filled by nerve C8. There is no C8 vertebra.

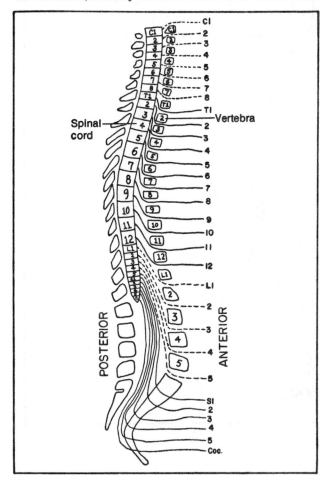

Figure 2-6

Note the normal curvature of the spine in figure 2-6. Excessive curvature of the small of the back (in the lumbar region) is termed **lordosis**; the buttocks appear to protrude. Excessive curvature in the thoracic region (hump-back) is termed **kyphosis**. Lateral bending of the spine (which cannot be seen in the lateral view of fig. 2-6) is **scoliosis**.

Fig 2-7. A typical thoracic vertebra, viewed from above. It looks like a happy snowman (petting the head of a moose - namely the head of a rib).

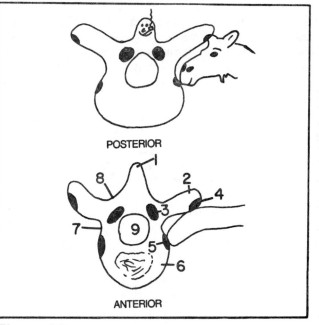

Figure 2-7

(1) spinous process (snowman's head)
(2) transverse process (arms)
(3) superior articular process (breast). The inferior articular processes are not seen ; they correspond to the snowman's scapulae). The articular processes connect adjacent vertebrae.
(4) facet - an area of contact between vertebra and rib (area where snowman's hand touches moose's head)
(5) another facet (snowman's front pocket) - an area of contact between vertebral body and rib. There is a third facet (not shown) posteriorly (snowman's rear pocket).
(6) body of vertebra (snowman's body). Intervertebral discs (not shown) separate the bodies of adjacent vertebrae.
(7) pedicle (snowman's waist)
(8) lamina (snowman's shoulder). Surgeons operating on the back sometimes remove the lamina in a procedure called a **laminectomy**. This procedure may be done to relieve pressure in the spinal canal following herniation of the center (**nucleus pulposus**) of a disc. The herniation may cause compression of a spinal nerve root. In the

lower back this is a common cause of "sciatica" ("lumbago").
(9) Vertebral canal - houses the spinal cord

The moose is trying to eat sugar lumps out of the front and rear pockets of the snowman while the snowman is petting it. These 3 points of contact are facets, areas of contact between transverse process and rib, and body and rib. Actually, the analogy is not complete, as the pockets which the moose is sniffing out do not belong to just one snowman but are the front pocket of one snowman and the rear pocket of the snowman immediately above it (fig. 2-8). There is no illicit contact between the snowmen as their bodies are separated by discs. These are cartilage-connective tissue cushions that allow flexibility in spinal movements. The articulations between vertebrae occur through the snowmen's breasts (breast to scapula) and bodies (with intervening discs).

Fig. 2-8. Vertebrae, lateral view (snowmen stacked up).

(1,2,3) facets for rib
(4) intervertebral foramen - note that the snowman has a steeply arched back. The spinal nerves exit on each side through the small of the back (intervertebral foramina).
(5) inferior articular process
(6) superior articular process
(7) pedicle
(8) lamina
(9) transverse process
(10) spinous process
(11) body

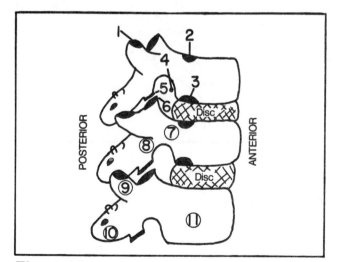

Figure 2-8

Although the above description is typical for a thoracic vertebra, other vertebrae differ slightly. The facets (pockets and petting area of the hand) are restricted to thoracic vertebrae as there are only 12 (thoracic) ribs.

Cervical snowmen have only one pocket on the right and left sides and the pocket has a hole in it (**transverse foramen**) to transmit the vertebral artery (fig. 2-9). Lumbar vertebrae have no such holes and no facets. They are just big, fat **lumbering** vertebrae. Lumbar spines don't have the steep, downward angulation of the thoracic vertebrae.

Fig. 2-9. Cervical, thoracic, and lumbar vertebrae.

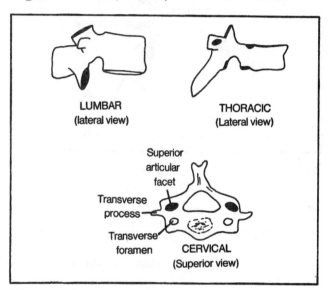

Figure 2-9

Fig. 2-10. The atlas and axis (superior views). These two cervical vertebrae deserve special mention. The atlas articulates with the skull and has no body. The axis bone below it, instead, contains the body (**dens**, or **odontoid process**) that properly should belong to the atlas. The dens fits into the atlas. The dens allows rotation of the atlas (and thus the head), as in shaking the head "no". The atlas articulates with the skull to allow anterior-posterior nodding of the head (shaking the head "yes").

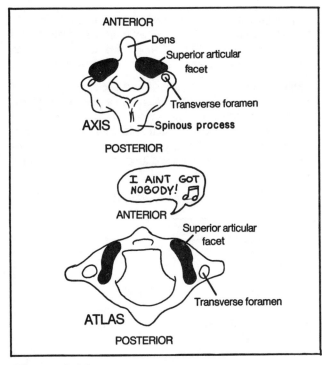

Figure 2-10

Fig. 2-11. The sacrum.

(1) sacral promontory - an important landmark in radiologic measurements
(2) pelvic (anterior, or ventral) sacral foramina
(3) coccyx
(4) sacral canal
(5) dorsal (posterior) sacral foramina
(6) sacral hiatus

The sacrum represents a developmental fusion of vertebrae. It attaches to the coccyx and looks like a butterfly. There is a peculiarity about the holes in the sacrum which will be elaborated upon in the chapter on the nervous system. Apparently, on each side of the body, each cervical, thoracic, or lumbar vertebra is matched with but one corresponding intervertebral foramen, through which one spinal nerve exits. Each section of the sacrum, however, has two associated foramina on each side, a posterior and an anterior sacral foramen. For now, though, don't let this little matter bother you. Pelvic nerves S1-S4 exit through the four sets of sacral foramina. In addition, paired nerves for S5 and the coccygeal nerve exit through the midline sacral hiatus. There are variations in the number of foramina.

Spondylolysis and Spondylolisthesis (fig. 2-12)

Spondylolysis and spondylolisthesis are conditions which, like a herniated disc, may contribute to low back pain. In **spondylolysis** there is a defect in the vertebral lamina. The snowman's head (spinous process), shoulders (laminae) and scapulae (inferior articular processes) separate from the rest of the vertebra. Spondylolysis may precede **spondylolisthesis** in which the two segments separate even more widely, the anterior segment slipping anteriorly (fig. 2-12). Although there may be back pain there commonly are not other significant neurologic defects, as the posterior aspect of the vertebra remains in place. Moreover, the vertebral canal is widened rather than narrowed. Spondylolisthesis may result in difficult delivery due to altered pelvic proportions.

Fig. 2-12. Spondylolisthesis

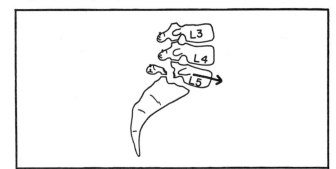

Figure 2-12

Fig. 2-13. The Pelvis

(1) Sacrum
(2) ilium
(3) ischium
(4) ischial tuberosity - the area that one sits on
(5) pubis
(6) symphysis pubis - a fibroelastic connection between the two pubic bones
(7) arcuate line
(8) pubic tubercle - the medial attachment point of the inguinal ligament (fig. 4-28)

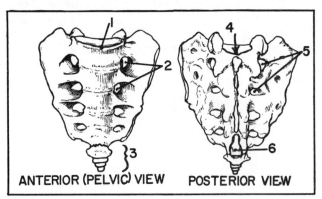

Figure 2-11

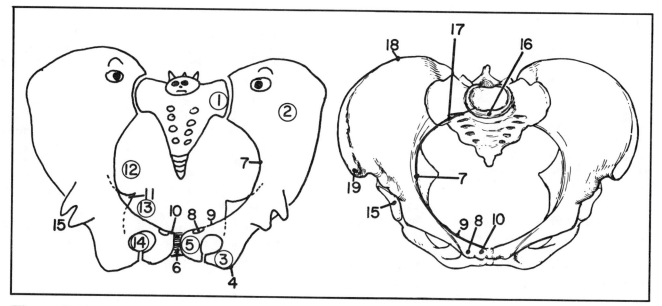

Figure 2-13

(9) pecten pubis - continuation of the arcuate line to the pubic tubercle

(10) pubic crest - connects pubic tubercle to symphysis pubis (the **rectus abdominis** muscle attaches here)

(11) ischial spine - separates greater and lesser sciatic notches

(12) greater sciatic notch - the sciatic nerve goes through here along with a lot of other things

(13) lesser sciatic notch

(14) obturator foramen

(15) acetabulum - the head of the femur fits in here

(16) sacral promontory

(17) terminal line (linea terminalis, the pelvic brim) -runs from sacral promontory to arcuate line to pecten pubis separates the **"false" pelvis** (above the terminal line) from the **true pelvis** (below the terminal line)

(18) iliac crest - is easily palpable because no muscles or tendons cross it

(19) anterior superior iliac spine - the inguinal ligament (fig. 12-19) stretches from here to the pubic tubercle.

The pelvis resembles two fish eating a butterfly. The butterfly is the sacrum - fig. 2-11). The bodies of the fish are the bodies of the ilium. Each fish has 2 tail fins. The "superior" fin of each fish is touching the "superior" tail fin of the other (in the region of the symphysis pubis). The superior tail fin is the pubis. The inferior tail fin is the ischium. In adults all these bones are fused. There is a back fin (the acetabulum). Attached to the inferior tail fin (ischium) within the pelvis is the ischial spine. The chin of the fish is the arcuate line. The true pelvis lies between the two fish below the level of the arcuate lines. The false pelvis lies above the arcuate lines and includes the area covering the fishes' faces and bodies.

Bones Of the Lower Extremity

Fig. 2-14. The femur, tibia, and fibula.

(1) head

(2) neck - commonly fractured, especially in women with postmenopausal **osteoporosis** where there is excessive bone resorption. Following an injury, outward rotation and shortening of the lower extremity suggests a fracture in this region

(3) greater trochanter

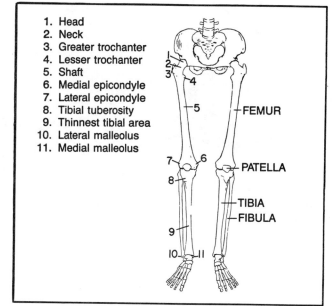

1. Head
2. Neck
3. Greater trochanter
4. Lesser trochanter
5. Shaft
6. Medial epicondyle
7. Lateral epicondyle
8. Tibial tuberosity
9. Thinnest tibial area
10. Lateral malleolus
11. Medial malleolus

Figure 2-14

(4) lesser trochanter

(5) shaft

(6) medial epicondyle

(7) lateral epicondyle

(8) tibial tuberosity (patellar ligament attaches here)

(9) thinnest portion of the tibia - a common site of fracture

(10) lateral malleolus of fibula - may fracture following forceful inversion or eversion of the foot

(11) medial malleolus of tibia - may be pulled off in a sprained ankle in which there is eversion of the foot (fig. 3-14)

The tibia is an important weight bearing bone. The fibula does not function in weight bearing, but mainly acts as a site for muscle attachment. One can even remove a section of fibula for purposes of obtaining a bone graft, without significant decrease in function of the lower extremity.

Fig. 2-15. Bones of the (right) foot. The bones of the foot are dominated by the TALUS, who insists from his dominant position on a calcified rock (calcaneous) that he is the TALLEST (talus) of all the foot bones. Most of the tarsal bones disagree with this.

The NAVICULAR replies "NEVER". The 3 CUNEIFORMS say "COULDN'T BE". Only the CUBOID bone, a real square, occupying an inferior position, looks up at the threatening talus and says "could be". The big bully talus is finally taught a lesson, as the tibia steps down decisively on it (the fibula, remember, is not a weight-bearing bone). No one likes a bully, and no muscle attaches to the talus.

The second metatarsal is particularly prone to **"march" fractures**, which occur when persons who are not in condition walk or run excessively.

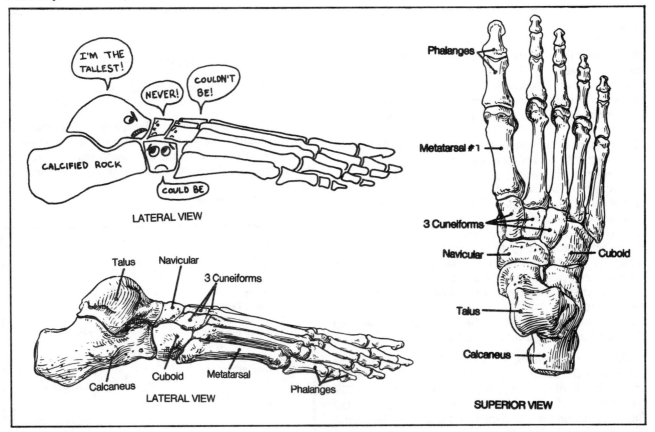

Figure 2-15

Fig. 2-16. **Hallux valgus** (bent first metatarsophalangeal joint) - may be caused by excessively pointy shoes. This same poor joint is also the one affected by inflammation in **gout.**

Cranial Bones

Fig. 2-17. The cranial bones.

(1) frontal
(2) sphenoid
(3) nasal
(4) lacrimal
(5) zygomatic
(6) ethmoid
(7) inferior nasal conch
(8) vomer
(9) maxilla
(10) mandible

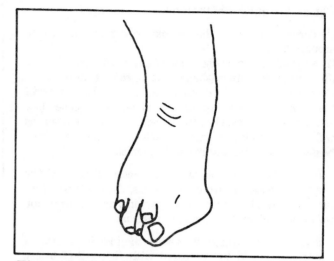

Figure 2-16

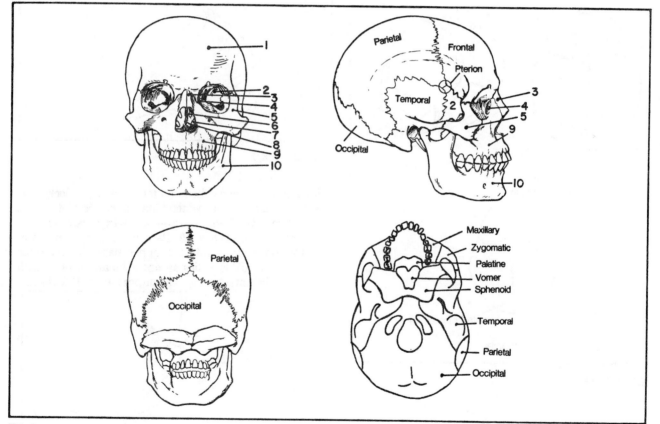

Figure 2-17

Fig. 2-18. Sutures of the Skull.

(1) coronal suture (shaped like **a corona**, or crown, around the head)
(2) sagittal suture (in the plane of a sagittal section)
(3) lambdoidal suture (shaped like a greek lambda)
(4) pterion - an "H"-shaped confluence of the sutures of the frontal, parietal, sphenoid and temporal bones (see also fig. 2-17. It overlies the speech area of the brain as well as the middle meningeal artery. These areas may be damaged with fractures of the pterion.

In infants the sutures are incompletely closed and the brain is exposed in two critical areas, the **posterior fontanelle** (closes by 2 months postpartum) and the **anterior fontanelle** (closes by 18 months postpartum).

Figures 2-19 through 2-33 show the individual cranial bones and should be compared with figure 2-17 to see them in proper perspective.

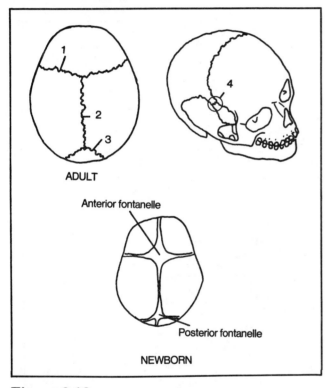

Figure 2-18

Fig. 2-19. Frontal bone (frontal view).

(1) frontal sinus - paired, roughly triangular in shape, involved in frontal sinus headaches.
(2) supraorbital foramen (or notch in some individuals). Surgeons sometimes anesthetize the supraorbital nerve where it exits this foramen, to achieve upper lid anesthesia in eyelid surgery.

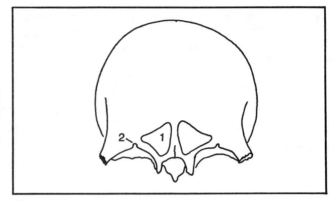

Figure 2-19

Fig. 2-20. Right parietal bone (lateral view). Not too interesting looking; one on each side of the head.

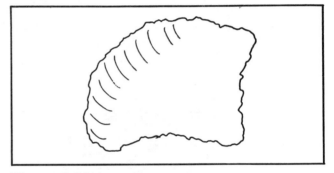

Figure 2-20

Fig. 2-21. Sphenoid bone (anterior view). It looks (vaguely) like a flying pterydactyl (has pterygoid plates -8,9). It has two sets of wings - a **lesser wing** above (1) and a **greater wing** below (2). Its eyes (3) are openings of the **sphenoid sinus** into the upper nasal cavity. Near each of the bird's eyes lies an **optic foramen** (4) (which transmits the optic nerve). The **superior orbital fissure**

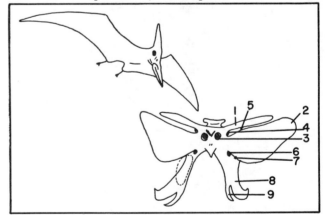

Figure 2-21

12

(5) (which transmits cranial nerves 3,4,5, and 6) is the superior armpit between the wings. The **round foramen**, or foramen rotundum (6), (which transmits the second, maxillary, branch of cranial nerve 5), and **inferior orbital fissures** (7) form the inferior armpit below the greater wing. The feet have two giant claws - the **lateral (8)** and **medial (9) pterygoid plates**. The **pterygopalatine fossa** (nerves subserving lacrimation and sensation to the upper teeth pass through it) is a space approximately outlined by the dotted lines in front of the pterydactyl's knee. If the knee were to kick the object directly in front of it, it would hit the maxillary bone. For further orientation, compare with figures 2-22 through 2-25, and 2-29.

Fig. 2-22. The sphenoid bone (dotted lines) in relief against other skull bones, anterior view.

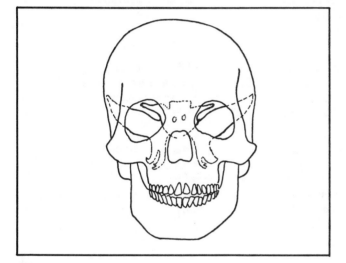

Figure 2-22

Fig. 2-23. Dorsal view of the sphenoid bone (shaded area) within the skull.

(1) lesser wing of sphenoid
(2) greater wing of sphenoid
(3) optic foramen - transmits optic nerve
(4) foramen rotundum - transmits maxillary branch (V2) of trigeminal nerve
(5) foramen ovale - transmits mandibular (V3) branch of trigeminal nerve
(6) foramen spinosum - transmits middle meningeal artery
(7) posterior clinoid process
(8) Sella turcica (fossa for the pituitary gland). This corresponds to a dent in the head of the pterydactyl in figure 2-21.

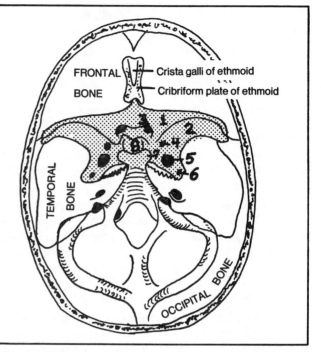

Figure 2-23

Fig. 2-24. How to get to the dark recesses of the hidden and mysterious pterygopalatine fossa. Ease your way (arrow) into the narrow canyon that lies between the lateral pterygoid plate and maxillary bone. You will then be wedged into this fossa. Alternatively, as depicted in figure 2-29, crawl into the orbit and lean over to look into the abyss of the inferior orbital fissure. If you fall in, you will fall directly into the pterygopalatine fossa.

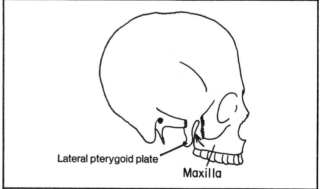

Figure 2-24

Fig. 2-25. Sagittal view of the skull, highlighting the sphenoid bone.

(1) sphenoid sinus - connects with the nasal passages
(2) pituitary fossa (sella turcica) - houses the pituitary gland; enlargement of this fossa (as by a pituitary tumor) may easily be detected on x-ray
(3) posterior clinoid process - an important landmark in radiology. It may appear eroded with increased intracranial pressure. Note the closeness of the sphenoid bone to the nasal passages. Surgeons may remove pituitary tumors through the nasal passages by cutting through the sphenoid bone.

Fig. 2-26. The ethmoid bone (shaded in B). From above (A) it looks like a tank (see also fig. 2-23) but it is an incredibly fragile bone. Note (B) its fragile lateral wall, which forms part of the lateral wall of the orbit (see also fig. 2-29). In orbital trauma the ethmoid bone may fracture and allow air to enter the orbit from the nasal cavities. The bone contains many delicate air cells.

Olfactory nerve fibers extend from the nasal cavity through the roof of the ethmoid bone (through holes in the cribriform plate) to the interior of the skull, where olfactory sensation is relayed to the brain. A fracture through the cribriform plate may cause cerebrospinal fluid seepage into the nose.

Each half of the nose has three nasal conchae. The ethmoid bone contains the upper two. The inferior nasal conch is a separate bone, shaped something like an inverted "U".

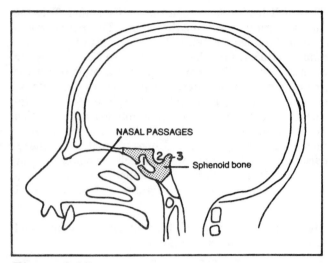

Figure 2-25

Fig. 2-27. The temporal bone. This jagged bone looks like the head of a chicken when viewed laterally. It looks like the head of a whale (petrous portion of the temporal bone) when viewed superiorly.

(1) squamous portion of temporal bone (chicken's comb)
(2) temporomandibular fossa (chicken's forehead) - the articular condyle of the mandible (fig. 2-32) fits into the chicken's indented forehead
(3) external acoustic meatus (chicken's eye) - the entrance to the external acoustic canal

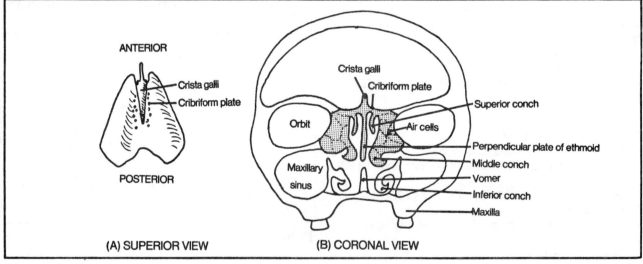

Figure 2-26

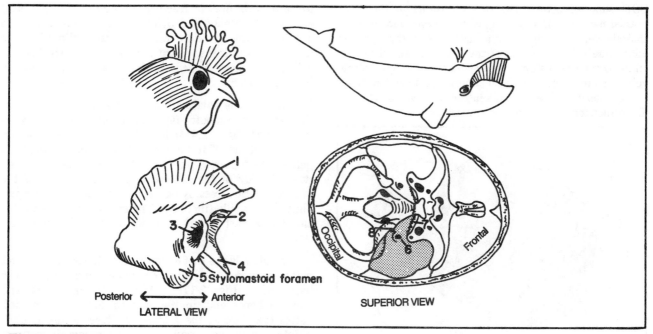

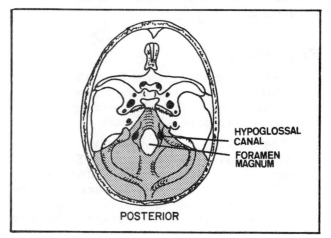

Figure 2-27

(4) styloid process (chicken's beak) - a muscle attachment site

(5) mastoid process (chicken's wattles)

(6) internal acoustic meatus (whale's eye) - cranial nerves 7 and 8 pass through here

(7) foramen lacerum (whales's teeth) - the carotid artery passes nearby, although not through, here

(8) jugular foramen (whale's spout) - the internal jugular vein passes through here. as well as cranial nerves 9, 10, and 11

Baby chickens are born with good hearing and balance and the temporal bone houses the auditory and vestibular apparatuses (cranial nerve 8). The facial nerve also passes through this bone and exits at the **stylomastoid foramen** (which obviously lies between the styloid and mastoid processes). Surgeons sometimes inject the facial nerve near this foramen to achieve paralysis of the eye muscles during eye surgery.

Fig. 2-28. The occipital bone. It has a big hole in it - the **foramen magnum**, where the spinal cord enters the skull and becomes the brain stem. It is the last (most posterior) bone of the head and the last cranial nerve (cranial nerve 12) goes through its **hypoglossal canal.**

If you wish to remember all the foramina for the cranial nerves, you will know you have holes in your head if you study the TOES:

Temporal bone (has foramina for 5 cranial nerves - 7, 8, 9, 10, 11)
Occipital (1 nerve - CN 12)

Ethmoid (1 nerve - CN1)
Sphenoid (5 nerves - CNs 2, 3, 4, 5, 6)

HYPOGLOSSAL CANAL
FORAMEN MAGNUM
POSTERIOR

Figure 2-28

Fig. 2-29. The orbital bones.

I.O., insertion of inferior oblique muscle
IOF, inferior orbital fissure - an observer looking down into this fissure would see the pterygopalatine fossa.
Lacr., lacrimal bone
O, optic canal - transmits the optic nerve and ophthalmic artery
P, palatine bone
S.g., sphenoid - greater wing
S.l., sphenoid - lesser wing

SOF, superior orbital fissure - transmits to the orbit cranial nerves 3, 4, 5, 6, sympathetic nerves and the ophthalmic vein (the ophthalmic artery goes through the optic canal). Enlargement of the superior orbital fissure on x-ray may be a sign of enlargement of the ophthalmic vein. This may occur with an abnormal connection (fistula) between carotid artery and intracranial veins.
T, trochlea for superior oblique muscle

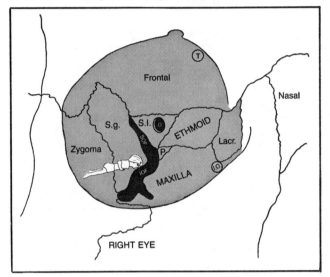

Figure 2-29

Facial Bones

The lacrimal bone looks sort of like nothing, kind of flaky - like the nasal bone (fig. 2-29). The vomer (fig. 2-26), which forms part of the midline nasal septum, is also thin and flaky and doesn't look like much.

Fig. 2-30. The zygomatic and maxillary bones (compare with fig. 2-17). The zygomatic bone is X-shaped and commonly is involved in lateral wall orbital fractures (fig. 2-29). The maxillary bone houses the **maxillary sinus**. Orbital floor fractures may easily break through into the maxillary sinus as may dental infections of the upper

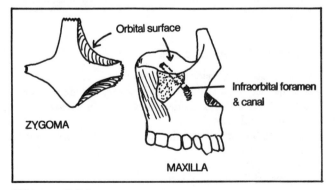

Figure 2-30

teeth. An orbital floor fracture through the **infraorbital canal** will damage the **infraorbital nerve** and cause local anesthesia of skin under the lower eyelid. Surgeons may inject the **infraorbital foramen** to anesthetize the infraorbital nerve to achieve anesthesia of the lower lid in eyelid surgery.

Fig. 2-31. The palatine bone. It strangely lies partly in the back of the mouth (fig. 2-17, basal view) and partly in the orbit (fig. 2-29). It is a "J" shaped bone. The bottom hook of the "J" forms the posterior aspect of the hard palate. The top cross bar of the "J" lies in the orbit. The straight body of the "J" forms the medial wall of the pterygopalatine fossa, lying between the lateral pterygoid plate and maxilla (fig. 2-24). The straight body of the "J" also forms part of the posterior wall of the maxillary sinus.

The palate is the roof of the mouth. It consists of a soft palate (posterior 1/3) and a hard palate (anterior 2/3). The hard palate (fig. 2-17) is formed by maxillary and palatine bones and, in the embryo, has right and left halves that fuse at the midline. Failure to fuse is **cleft palate**, which may require surgical repair. Sometimes the midline junction is especially protuberant (**torus palatinus**). This benign condition should not be confused with a tumor.

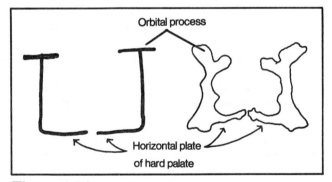

Figure 2-31

Fig. 2-32. The Mandible. The mandible looks like **a** mandible.

(1) articular condyle - fits into the temporal bone (i.e., into the chicken's forehead), forming the hinge around which the jaw swings. Irregularities of this joint may cause pain on moving the jaw (the **temporomandibular joint syndrome**)
(2) coronoid process. The temporalis muscle (fig. 4-59) attaches here and assists in jaw closing when it pulls up
(3) ramus - hugged by the parotid gland (fig. 9-2)
(4) angle of the mandible - may determine whether someone looks rugged or not
(5) entrance to the mandibular canal - transmits the **inferior alveolar nerve** (a branch of the mandibular

branch of the trigeminal nerve), which innervates the lower teeth.

(6) mandibular canal

(7) mental foramen - the inferior alveolar nerve continues on through this foramen as the **mental nerve**, which supplies chin skin. Hence, after local dental anesthesia of the inferior alveolar nerve, the patient may also experience numbness of the chin and lower lip on that side.

If it were not for the mandible, dentists would earn half as much, as it contains half of the teeth.

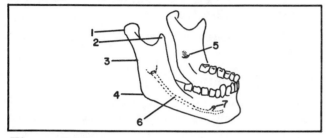

Figure 2-32

Fig. 2-33. The hyoid bone - looks like a miniature mandible with two incisor teeth (lesser horns); an important muscle attachment site.

Development Of Bone

Most bones initially are **cartilaginous** and then change to bone. However, the bones of the skull, face, and clavicle are **membranous**, developing from embryonic connective tissue, without any intermediate cartilaginous phase.

Bone is produced by cells called **osteoBlasts** (the Builders). The form of growing bone is maintained by cells called **osteoClasts** (the sCulptors).

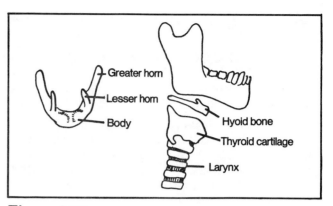

Figure 2-33

Fig. 2-34. Development of cartilaginous bone. In fetal development, bone is cartilaginous and its midshaft develops a "**primary**" **ossification center** which produces bone. Later on, generally between birth and puberty, bone is also produced by "**secondary**" **ossification centers** in the **epiphyses** (ends) of these bones. Growth of the cartilaginous bones procedes largely by the proliferation of new cartilage along with replacement of old cartilage with bone. Eventually, cartilage of the shaft is replaced by bone. What remains of the cartilage is a cartilaginous **epiphyseal plate** separating epiphysis and shaft. There is also cartilage at the very tips of the bone, where the bone is exposed to the joint cavity.

Continued elongation of bone depends on the presence of the epiphyseal plates which begin to disappear at puberty. Bone elongation stops with fusion of the epiphysis and shaft. The **periosteum**, which lines the bones also produces new bone, enabling the developing bone to thicken. The periosteum plays a major role in the repair of bone following fracture. The smooth, glistening cartil-

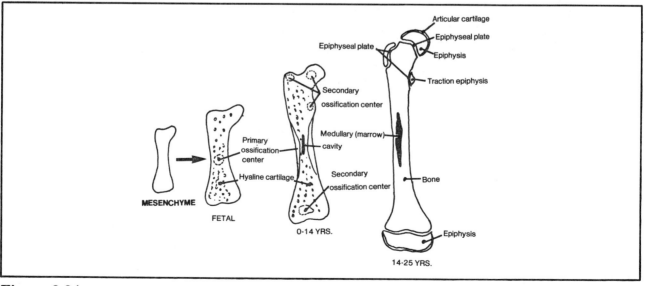

Figure 2-34

age, at those ends of bone that lie in a joint space, remains throughout life.

Membranous bones develop more simply, from primary ossification centers, in the absence of cartilage or "secondary" ossification centers. The bone begins with connective tissue which subsequently ossifies.

Cleidocranial dysostosis is a congenital condition in which membranous bone develops abnormally. There may be deformity of the cranial bones and absent clavicles. **Achondroplasia** is a disorder of cartilaginous bones, which do not grow in length, resulting in a dwarf with normal facial features.

The developmental age of an individual can often be determined radiologically by the stage of bone calcification. For instance, the clavicle is the first bone to calcify in the fetus, beginning about the 5th week of fetal development. A fetus is considered old enough to be full term (sometimes a medico-legal issue) if it has a secondary growth center in both the distal femur and proximal tibia. In children, the most useful area to examine to determine developmental age is the wrists and hands. Epiphyseal development in girls generally occurs somewhat in advance of that in boys.

The bones of young children are less brittle than those of adults. However, young bones do have epiphyseal growth plates that may slip out of alignment with the shaft (e.g., **slipped epiphysis** of the femoral or radial head). In children the epiphyseal region may also be more susceptible to vascular deprivation. Normally, the shaft receives its blood supply from an artery which pierces the shaft and then divides to supply both ends of the bone up to the epiphyseal plates. The epiphysis receives a separate supply from arteries of the joint capsule. The two blood supplies do not normally anastomose until fusion between epiphysis and shaft occurs. Before then, the epiphysis may be particularly susceptible to vascular compromise. E.g., in **Legg-Perthes disease** (usually occurring at ages 3-6), there is degeneration of the epiphysis of the head of the femur. In **Osgood-Schlatter disease** (usually ages 10-11) degeneration occurs in the epiphysis of the tibial tuberosity.

Fracture through an epiphysis may retard bone growth.

Sesamoid Bones

Sesamoid bones are floating bones that occur within tendons, generally near a joint or in a region where tendons turn acutely around a bony prominence. As they are usually symmetrical, they may be distinguished from fractures by x-rays that compare both sides of the body. They commonly occur in the hand and foot. The largest sesamoid bone, however, is the patella (knee cap). The **fabella** is a smaller sesamoid bone, lying behind the lateral femoral condyle of the knee in the gastrocnemius tendon. A sesamoid bone may also overly the greater trochanter of the femur.

CHAPTER 3.
SKELETAL LIGAMENTS, CARTILAGE, AND TEETH

Skeletal ligaments

Skeletal ligaments are tough fibrous bands that bind bones together and stabilize the joints. Freely-mobile joints also contain a synovial cavity. This cavity is lined inside by **synovial membrane**, which produces joint fluid. The fluid reduces the friction of the moving bones. The synovial membrane is surrounded by a **fibrous joint capsule** which, in key places, is reinforced by ligaments. In the knee, for instance, one naturally expects to find the joint capsule strengthened by ligaments on the medial and lateral sides of the knee - to prevent medial or lateral bending. The anterior and posterior aspects of the fibrous knee capsule are less strong, to allow greater flexibility in bending. Some joints, such as the knee, also contain **fibrocartilaginous discs** to act as cushions between the bones.

Fig. 3-1. A synovial joint.

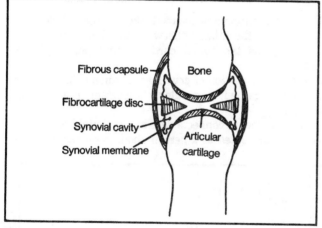

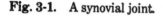

Figure 3-1

Fig. 3-2. Lateral ligament of the jaw. The lateral ligament does a good job of preventing the temporomandibular joint from dislocating backwards after getting punched in the jaw. However, the joint not uncommonly dislocates anteriorly (such as after becoming bored with reading about ligaments and issuing a giant yawn). If this occurs, you may reduce the dislocation by applying your thumb to the patient's lower molars and pressing down and backwards, pushing the mandible back into place.

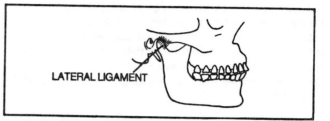

Figure 3-2

Fig. 3-3. Ligaments of the shoulder girdle:

(1) interclavicular ligament - connects the clavicles; homologous with the wishbone in chickens
(2) sternoclavicular ligament - contributes to stability of the sternoclavicular joint; helps prevent the medial end of the clavicle from flipping up and out of place when one walks with a heavy suitcase
(3) coracoclavicular ligament - superstrong ligament. If ruptured, e.g. following a severe fall on the shoulder in football, the clavicle may dislocate over the acromion (a **shoulder separation**). The dislocation is even worse if the **acromioclavicular ligament**
(4) also tears. The scapula then drops down and the lateral end of the clavicle protrudes prominently.

Rupture of the **coracoacromial ligament**
(5) does not result in such displacement as this ligament connects two parts of the same bone (the scapula). However, the coracoacromial ligament is important in blocking the head of the humerus from dislocating superiorly.
(6) the fibrous capsule of the shoulder joint - is weakest inferiorly, where dislocation of the head of the humerus is most likely to occur.

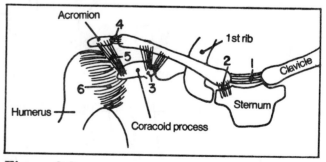

Figure 3-3

Fig. 3-4. The elbow joint.

(1) radial collateral ligament - binds radius to humerus
(2) ulnar collateral ligament - binds ulna to humerus
(3) annular ligament - binds radius to ulna. Small children who are picked up rapidly by the hand may experience a **"pulled elbow"** in which the annular ligament tears and the head of the radius dislocates. Correction of this may often be obtained immediately by supinating the forearm, as shown in the same figure. In "SUPination" the hand is turned over in such manner that it can hold a cup of "SOUP".

Fig. 3-5. Vertebral ligaments. Various ligaments participate in stabilizing the stack of vertebral snowmen (see also fig 3-6):

(1) supraspinous ligament - interconnects the head tops of the snowmen (figs. 2-7,2-8). In the neck, this ligament runs right onto the back of the skull as the **ligamentum nuchae**
(2) ligamentum flavum (yellow ligament)
(3) posterior longitudinal ligament - prevents hyperflexion of the vertebral column; continues to the skull as the **tectorial membrane**
(4) ANTerior longitudinal ligament - lies in the ground under the snowman's body - where the "ANTS" are. It prevents overextension of the vertebral column and may tear during a whiplash injury. It continues to the skull as the **anterior atlantooccipital membrane.**

Not shown in figure 3-5 are all the **intertransverse ligaments** (connect the hands of the snowmen) and **interspinous ligaments** (connect the face of one snowman with the back of the head of the snowman on top) and

ligaments connecting the upper extremity of the snowman with the ribs (moose being petted in fig. 2-7). In a spinal tap, the needle will penetrate the supraspinous ligament, interspinous ligament and yellow ligament.

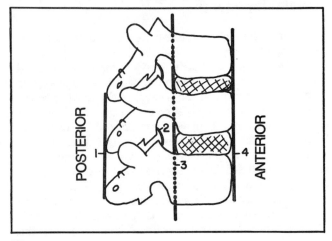

Figure 3-5

Fig. 3-6. Major ligaments of the atlas and axis.

(1) cruciform ligament (shaped like a cross) - holds the dens against the atlas. Otherwise, the dens could move and compress the spinal cord, causing paraplegia or death
(2) transverse portion of cruciform ligament - part of the cruciform ligament particularly important in anchoring the dens
(3) alar ligament - connects dens and occipital bone
(4) dens - presented as a person about to be hanged. The

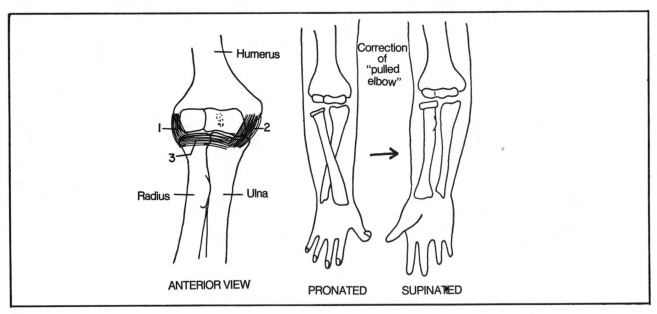

Figure 3-4

blindfold is the transverse part of the cruciform ligament and the hair standing on end is the alar ligament. The transverse ligament may rupture during a hanging, resulting in death, due to indentation of the cord by the dens. The transverse ligament is strong, however, and the more usual cause of death is a combination of fractures and dislocations of the 2nd, 3rd, and/or 4th cervical vertebrae.

If you were situated inside the spinal canal, you would not see the cruciform ligament. It is covered by the **tectorial membrane** (not shown in fig. 3-6) which is the continuation of the **posterior longitudinal ligament up to the skull** (fig. 3-5).

Fig. 3-7. Ligaments of the pelvis.

(1) iliolumbar ligament - attaches transverse process of L5 vertebra to ilium; helps prevent L5 vertebra from slipping forward (spondylolisthesis)

(2) sacroiliac ligaments - helps maintain union between sacrum and ilium

(3) sacrotuberous ligament - stretches from sacrum to ischial tuberosity; helps to maintain a stable relationship between sacrum and ischium

(4) sacrospinous ligament - helps to maintain a stable relationship between sacrum and ischium. During pregnancy and delivery these above-mentioned ligaments, as well as the symphysis pubis, relax. This allows greater mobility of the associated joints and more room for the fetus to be delivered.

(5) greater sciatic foramen - the "doorway to the pelvis". It is the exit through which the sciatic nerve passes to the lower extremity

(6) lesser sciatic foramen - the "doorway to the perineum". It is the passageway for the pudendal nerve to the perineum (i.e., to the anal and urogenital triangles).

Fig. 3-8. The iliofemoral ligament of the hip joint. A fibrous capsule encloses the hip joint. Thickened areas of

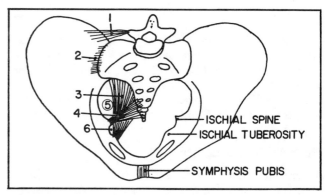

Figure 3-7

the fibrous capsule (such as the iliofemoral ligament) provide added strength. The iliofemoral ligament prevents the hip from overextending when standing. In fact, this ligament stands on its own two legs (it is shaped like an inverted "Y").

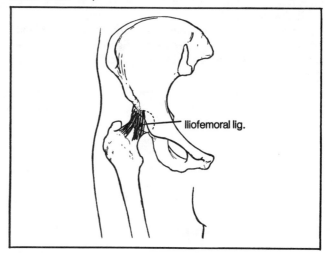

Figure 3-8

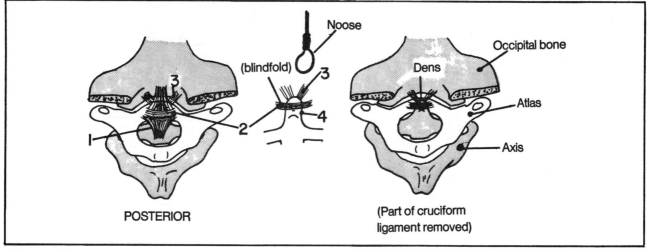

Figure 3-6

The Knee Joint

Unlike the humerus, which articulates with both the radius and ulna, the femur articulates with the tibia, but not the fibula (fig. 3-9). A fibrous capsule surrounds, protects and helps support the knee joint. A synovial membrane lines the inside of the fibrous capsule and produces joint fluid. The knee joint is the largest synovial joint of the body.

Fig. 3-9. Ligaments of the knee.

(1) patellar ligament - an extension of the quadriceps femoris muscle tendon. Tapping this ligament induces the **knee jerk reflex**

(2) tibial (medial) collateral ligament

(3) fibular (lateral) collateral ligament

(4) medial meniscus (the menisci are fibrocartilaginous cushions between femur and tibia)

(5) lateral meniscus

(6) anterior cruciate ligament - attaches to anterior (and medial) aspect of tibia

(7) posterior cruciate ligament - attaches to posterior (and lateral) aspect of tibia. The external ligaments of the knee are part of the fibrous joint capsule and help strengthen the joint

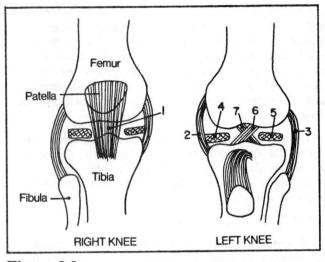

RIGHT KNEE   LEFT KNEE

Figure 3-9

Fig. 3-10. Damage to the **medial (tibial) collateral ligament** may occur with lateral blows to the knee. As this ligament attaches to the medial meniscus, both the medial collateral ligament and medial meniscus are commonly torn together. The anterior cruciate ligament may also tear in such injuries. The **lateral (fibular) col-**

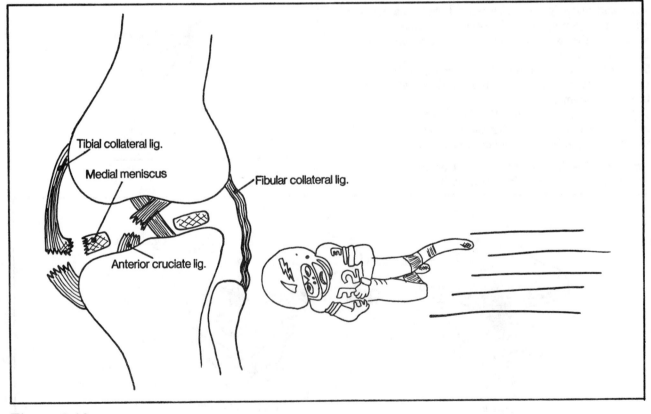

Figure 3-10

lateral ligament is not damaged often, as medial blows to the knee are uncommon. Note that the lateral collateral ligament does not attach to the lateral meniscus.

Tears of the **anterior cruciate ligament** can be diagnosed by noting that the tibia can be pulled forward (anterior drawer sign - fig. 3-11). Following tears of the **posterior cruciate ligament**, the tibia can be pulled backward, (posterior drawer sign).

Fig. 3-11. Anterior and posterior drawer signs.

Fig. 3-12. Knee injection and extractions of synovial fluid are commonly made lateral to the patella. Place your finger in the depression that lies lateral to the patella so that the finger simultaneously feels the patella, tibia, and femur. The needle is inserted into this site perpendicular to the skin. In figure 3-12, the dotted line indicates the extent of the joint cavity as seen anteriorly.

Fig. 3-13. A right ankle (schematic view). The posterior tibiofibular ligament may pull on the tibia during a severe injury, causing a fracture of the tibia at (C). A combined fracture of (A) (medial malleolus), (B) (lateral malleolus) and (C) (posterior part of the inferior tibial border) is a **trimalleolar fracture.**

Fig. 3-14. The medial and lateral ligaments of the ankle.

(A) The medial (deltoid) ligament is quite strong, binding the tibia to 3 bones, the calcaneus, navicular, and talus.

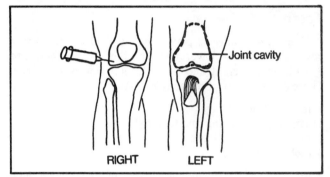

Figure 3-12

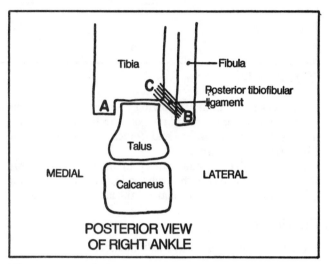

Figure 3-13

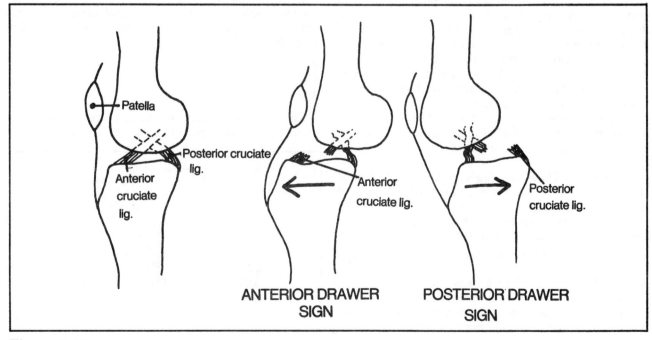

Figure 3-11

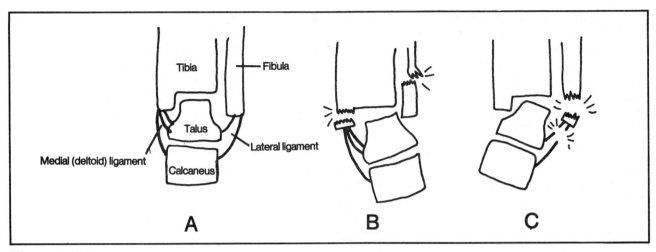

Figure 3-14

(B) In a twisting injury, involving eversion of the foot, the strong deltoid ligament might not tear but instead pulls off the medial malleolus and forces a fracture of the fibula (**Pott's fracture**).

(C) The relatively weak lateral ligament is commonly injured, particularly with inversion injuries, where it may tear (ankle sprain) and/or tear off the lateral malleolus.

Cartilage

Cartilaginous areas of the body include:

1. The firm framework of the ear and the lateral 1/3 of the external auditory canal.
2. Points of connection between the skull bones. Such cartilaginous areas generally disappear with age.
3. Part of the auditory (Eustachian) tube.
4. The firm areas of the nose, anterior to the nasal bone, including the anterior aspect of the nasal septum.
5. The larynx (including thyroid, cricoid, arytenoid, corniculate, cuneiform, and epiglottic cartilages) and the tracheal and bronchial rings (fig. 4-65, 4-67).
6. Costal cartilage at the anterior ends of the ribs.
7. Medial and lateral menisci of the knee. These are fibrocartilaginous plates in the knee between the femur and tibia (fig. 3-9).
8. Symphyses. These are fibrocartilaginous structures that connect bones and are not associated with a joint cavity. These include: intervertebral discs, the symphysis pubis, and the sternomanubrial symphysis (between manubrium and body of the sternum).
9. Cartilage capping the ends of bones in joint spaces (fig. 3-1).
10. Epiphyseal plates. This type of cartilage provides a growth center for developing bone (fig. 2-34). With development, the head (epiphysis) of the developing bone fuses with the diaphysis (shaft region) and growth stops.

Teeth

There are usually 32 permanent teeth, 16 in the upper jaw and 16 in the lower jaw. In other words there are 8 teeth on the right and 8 on the left side of each jaw. These 8 permanent teeth are (fig. 3-15):

2 incisors (for cutting)
1 canine (for piercing and tearing)
2 premolars (for grinding)
3 molars (grinding)

Fig. 3-15. The permanent teeth of the upper jaw. Shaded teeth do not have preceding milk teeth.

The 8 permanent teeth follow 5 temporary (milk, deciduous) teeth. Thus, the 3 permanent molars arise one time only and do not replace any milk teeth. The 5 milk teeth are:

2 incisors
1 canine
2 molars

Note that the 2 deciduous "molars" are pushed out by permanent **premolars** and not by permanent molars.

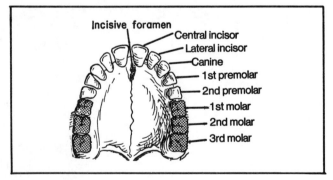

Figure 3-15

Fig. 3-16. A general scheme of the times of eruption of the teeth. The times shown are roughly rounded off to facilitate memory. In general, the lower teeth erupt before the upper teeth. The first temporary tooth erupts at about 6 months (generally a lower medial incisor). The first permanent tooth erupts at about 6 years (the "six-year first molar").

Tooth eruption occurs in approximately a medial to lateral sequence, although the canine is a little delayed in this regard.

Note that the **temporary** teeth erupt at approximately 4 month intervals (6, 10, 14, 18, and 22 months), beginning at 6 months and are replaced by their **corresponding permanent teeth** at yearly intervals, beginning at about 7-8 years (approximately 8, 9, 10, 11, and 12 years). In addition, the three solitary permanent molars erupt at approximately 6, 12, and 24 yrs. The first teeth to shed are usually the lower two medial incisors, at about 6 or 7 years. There may be as much as a year's delay between loss of a deciduous tooth and eruption of its corresponding permanent tooth.

AGE OF TOOTH ERUPTION

TEMPORARY TEETH	MEDIAL INCISOR	LATERAL INCISOR	IST MOLAR	CANINE	2ND MOLAR			
	6 mos.	10 mos.	14 mos.	18 mos.	22 mos.			

PERMANENT TEETH	MEDIAL INCISOR	LATERAL INCISOR	IST PREMOLAR	CANINE	2ND PREMOLAR	IST MOLAR	2ND MOLAR	3RD MOLAR
	7-8 yr.	9yr.	10 yr.	11 yr.	12 yr.	6 yr.	12 yr.	24 yr.

Figure 3-16

CHAPTER 4. THE MUSCULAR SYSTEM

It is helpful to remember, as a general principle of muscle anatomy, that **when two muscles overlap one another and run in parallel, the shorter muscle tends to lie deeper.** General patterns of muscle innervation are discussed in chapters 12-14 (Nervous System). Specific listings are given in the Appendix, page 173.

Muscles That Move the Shoulder

Fig. 4-1. Shoulder movement. Shoulder movement depends not only upon movement at the glenohumeral joint but also upon movement of the scapula. Elevation of the upper limb begins at the glenohumeral joint but is completed by rotation of the scapula. The logical classifications of muscles that would move the shoulder are:

A. muscles connecting trunk to scapula
B. muscles connecting scapula to humerus
C. muscles connecting trunk to humerus

Fig. 4-2. Muscles connecting the trunk to the right scapula: THE BULLFIGHT (Scene I). The bull (scapula) is being restrained by four hands (muscles connecting trunk to scapula). The torreador's cape is the trapezius. The four hands restraining the bull are the rhomboids (count as one hand), levator scapulae, pectoralis minor, and serratus anterior.

(1) trapezius (red cape) - adducts, rotates scapula. Su-

perior fibers elevate the shoulder; inferior fibers lower the shoulder.
(2) pectoralis minor - connects ribs 3, 4,and 5 with the coracoid process. Moves the scapula forward, and down.
(3) levator scapulae - elevates the scapula

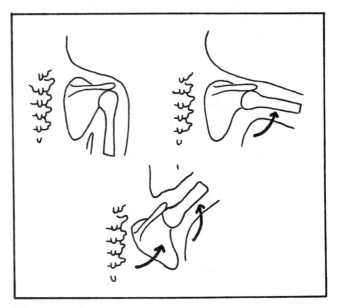

Figure 4-1

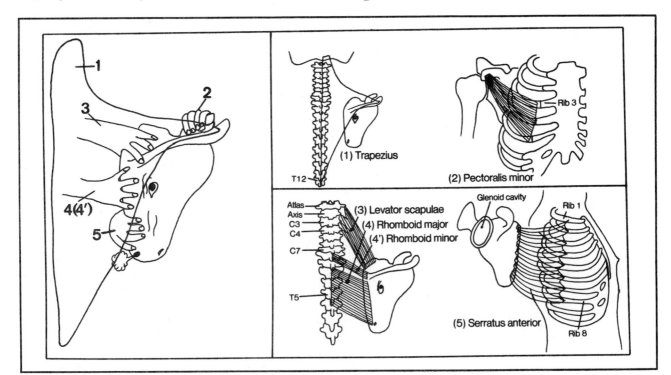

(1) Trapezius

(2) Pectoralis minor

Atlas
Axis
C3
C4
C7
T5

(3) Levator scapulae
(4) Rhomboid major
(4') Rhomboid minor

Glenoid cavity
Rib 1

(5) Serratus anterior
Rib 8

Figure 4-2

(4) rhomboids (major and minor) - retract and elevate the scapula

(5) serratus anterior - lies deep to the scapula; connects the upper 8 ribs with the medial border of the scapula. This muscle rotates the scapula so as to assist in abduction of the upper limb. It also moves the scapula forward to enable a boxer to knock out his opponent ("**the boxer's muscle**"). In addition, it holds the scapula close to the rib cage; paralysis of the nerve (long thoracic nerve) to this muscle results in "**scapular winging**" (fig. 4-3).

The trunk-to-scapula muscles, in appropriate combination, can move the scapula up, down, forward, or back, or rotate it clockwise or counterclockwise.

Fig. 4-3. Scapular winging on the right. The nerve to the serratus anterior muscle (long thoracic nerve) has

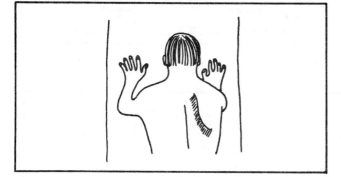

Figure 4-3

been damaged. The scapula protrudes on attempted forward pushing of the arms.

Fig. 4-4. BULLFIGHT - scene II. Muscles connecting right scapula to humerus. The bull is harnessed.

(1) deltoid (cape) - abducts arm. Note that the bull, after snagging the red cape (trapezius in fig. 4-2), has flung the cape (deltoid in present scene) over its head. The insertion of the trapezius muscle is almost the same as the origin of the deltoid muscle.

(2) supraspinatus (harness to scapular spine- abducts arm). The supraspinatus muscle initiates shoulder abduction; the deltoid continues abduction to the horizontal. Elevation above the horizontal is then carried out by the muscles that rotate the scapula.

(3) subscapularis (harness to anterior face) - rotates and adducts humerus

(4) infraspinatus (harness to posterior face) - rotates and adducts humerus

(5) teres minor (harness to chin) - rotates and adducts humerus

(6) teres major (harness to mouth) - rotates and adducts humerus

(7) coracobrachialis - does not quite fit in with the other muscles of this group as it is located predominantly in the arm. It is depicted in the same picture as the subscapularis (fig. 4-4), connecting coracoid process and medial aspect of the mid-humerus. It flexes and adducts the arm. The coracobrachilis is the only muscle that predominantly moves the shoulder while lying mainly in the arm.

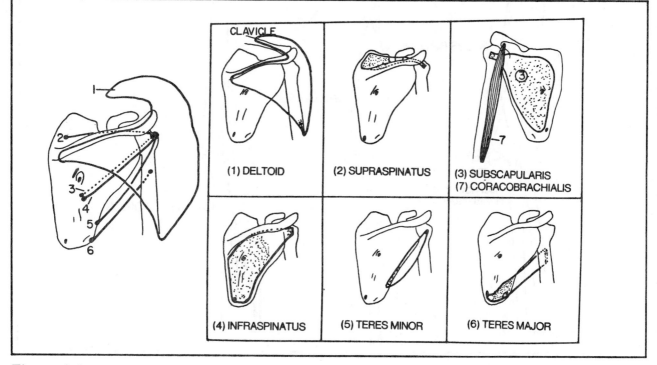

Figure 4-4

The **rotator cuff** is a grouping of muscles under the deltoid. The rotator cuff muscles are also called the "SITS" muscles (S-supraspinatus; I-infraspinatus; T-teres minor; S-subscapularis), as an athlete who injures the rotator cuff "SITS" out the game (if not the season). The tendons of the SITS muscles fuse with the capsule of the shoulder joint and help stabilize the shoulder joint.

Fig. 4-5. Muscles connecting the trunk and humerus. There are only two, but they are big ones, covering the front and back of the trunk. The pectoralis major and latissimus dorsi both adduct and medially rotate the humerus. Note that the lower edge of the pectoralis major muscle rests on rib 7 while the upper edge of the latissimus dorsi lies on vertebra T7.

Muscles Of the Upper Extremities

Fig. 4-6. Muscles predominantly flexing the elbow.

(1) biceps brachii - also a powerful supinator of the forearm, as it attaches to the radial tuberosity (fig. 4-6A)
(2) brachialis. Don't confuse the brachialis with the co-racobrachialis (fig. 4-4). The coracobrachialis has nothing to do with elbow flexion, even though it is situated mainly in the arm. The brachialis attaches to the coronoid process of the ulna. If the ulna resembles a wrench, the olecranon is the upper jaw and the coronoid process is the lower jaw of the wrench. Sometimes, extremely

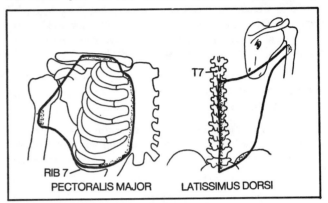

PECTORALIS MAJOR   LATISSIMUS DORSI

Figure 4-5

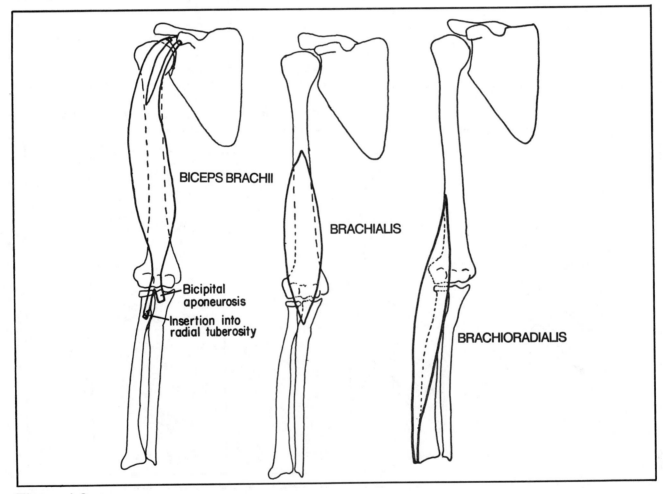

BICEPS BRACHII

Bicipital aponeurosis

Insertion into radial tuberosity

BRACHIALIS

BRACHIORADIALIS

Figure 4-6

vigorous contraction of the brachialis can avulse a portion of the coronoid process.

(3) brachioradialis - involved with elbow flexion but is mainly located in the forearm rather than the arm.

Fig. 4-6A. Function of the biceps (reproduced with permission, from Hoppenfeld, S.: Physical Examination Of the Spine and Extremities; Appleton-Century-Crofts, 1976, pg. 53). It acts as both flexor and supinator.

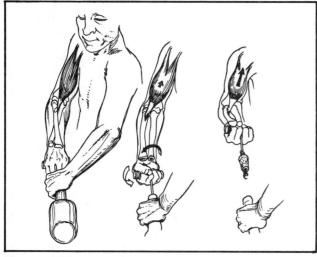

Figure 4-6A

Fig. 4-7. Muscles predominantly extending the elbow.

(1) triceps
(2) anconeus

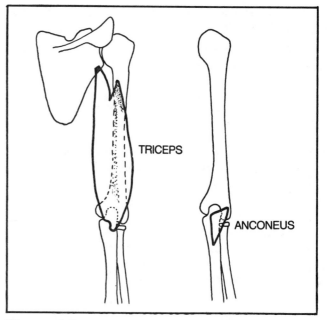

Figure 4-7

Muscles That Move the Wrist and Fingers

Muscles that move the wrist reside in the forearm. Muscles that move the fingers reside either in the forearm or are small muscles intrinsic to the hand. Feel your forearm as you flex your wrist and fingers. Note that, as a whole, the **flexors** tend to originate at, or in line with, the **medial epicondyle** of the humerus. In contrast, the

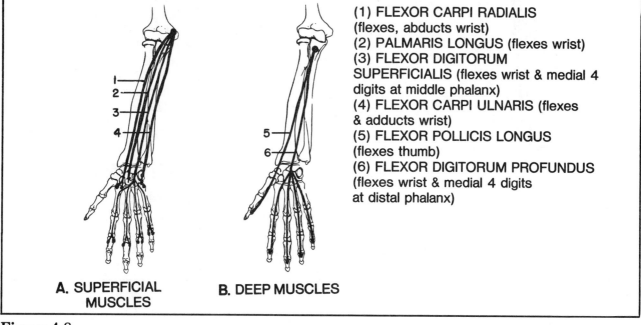

(1) FLEXOR CARPI RADIALIS
(flexes, abducts wrist)
(2) PALMARIS LONGUS (flexes wrist)
(3) FLEXOR DIGITORUM
SUPERFICIALIS (flexes wrist & medial 4
digits at middle phalanx)
(4) FLEXOR CARPI ULNARIS (flexes
& adducts wrist)
(5) FLEXOR POLLICIS LONGUS
(flexes thumb)
(6) FLEXOR DIGITORUM PROFUNDUS
(flexes wrist & medial 4 digits
at distal phalanx)

A. SUPERFICIAL MUSCLES

B. DEEP MUSCLES

Figure 4-8

extensors tend to connect (or align with) the **lateral epicondyle** of the humerus.

Fig. 4-8. Forearm muscles that flex the wrist and fingers. Note the alignment of the flexor muscles with the medial epicondyle. A, longer (more superficial) muscles; B, shorter (deeper) muscles. These figures are highly schematic; some of these muscles have multiple origins along their courses, including not only the medial epicondyle and ulna, but also the **interosseus membrane** (fig. 6-27) and **radius**. Such multiple origins particularly occur among the flexors (and extensors) of the fingers as these muscles are rather wide.

Note one peculiar exception to the rule of shorter muscles lying deeper. The flexor digitorum superficialis lies superficial to the flexor digitorum profundus, but its course on the hand is shorter than that of the profundus. The profundus tendon passes through a terminal split in the superficialis tendon to reach the last phalanx.

Fig. 4-9. Forearm muscles that extend the wrist and fingers. They tend to connect (or align) with the lateral epicondyle of the humerus. As for the flexors, there is a superficial and deep grouping; deeper muscles tend to be shorter, and originate in the forearm rather than at the lateral epicondyle. Note a certain similarity between the superficial extensors, as seen in this figure, and flexors in figure 4-8. Of course, the names differ; since there is no palm on the dorsum of the hand, the extensors can

not have a "palmaris" muscle. Instead, there is an extensor carpi radialis brevis muscle to extend the wrist.

The plan of the deeper extensors does not resemble that of the flexors. Rather, the tendons of the deep extensors appear pushed over toward the thumb. One of the muscles (abductor pollicis longus) got pushed over so far that it actually ended on the palmar surface of the hand and in fact is not even an extensor; it acts to abduct the thumb.

Fig. 4-10. The anatomical snuffbox, a pocket bounded by the extensor pollicis longus and extensor pollicis brevis.

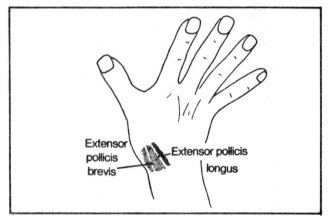

Figure 4-10

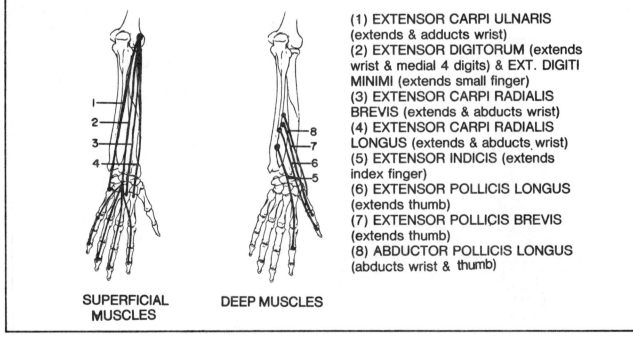

(1) EXTENSOR CARPI ULNARIS (extends & adducts wrist)
(2) EXTENSOR DIGITORUM (extends wrist & medial 4 digits) & EXT. DIGITI MINIMI (extends small finger)
(3) EXTENSOR CARPI RADIALIS BREVIS (extends & abducts wrist)
(4) EXTENSOR CARPI RADIALIS LONGUS (extends & abducts wrist)
(5) EXTENSOR INDICIS (extends index finger)
(6) EXTENSOR POLLICIS LONGUS (extends thumb)
(7) EXTENSOR POLLICIS BREVIS (extends thumb)
(8) ABDUCTOR POLLICIS LONGUS (abducts wrist & thumb)

SUPERFICIAL MUSCLES   DEEP MUSCLES

Figure 4-9

Fig. 4-11. Muscles of pronation and supination.

(1) pronator teres
(2) pronator quadratus
(3) supinator

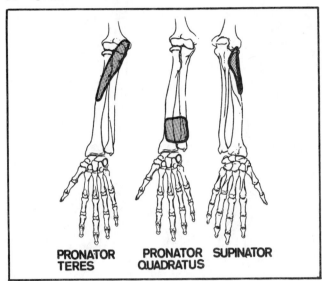

Figure 4-11

Fig. 4-12. Intrinsic muscles of the hand.

Eight muscles act on the thumb. Four of them originate in the forearm and have already been mentioned (abductor pollicis longus, flexor pollicis longus, extensor pollicis longus and extensor pollicis brevis). The other 4 are located in the hand (abductor pollicis brevis, flexor pollicis brevis, opponens pollicis, adductor pollicis).

The interossei and lumbricales (fig. 4-13) do not act on the thumb. How could they when all that room is taken up by the 8 thumb muscles? The dorsal interossei abduct the fingers. The palmar ones adduct. You can recall this by abducting your fingers. The fingers will have a slight tendency to extend at the metacarpophalangeal joint, reflecting the dorsal positioning of the abductor muscles. Adduction produces the opposite result, reflecting the palmar positioning of the palmar interossei.

Fig. 4-13. Lumbricales. These muscles act by flexing the metacarpophalangeal joint and extending the interphalangeal joints. Thus, they give the hand an "L" shape, as they flex and extend simultaneously. This action occurs because the lumbricales attach to the tendons of extensor digitorum and flexor digitorum profundus.

If you regard memorizing the intrinsic hand muscles as a chore, you may be interested to know that if you learn them you will have little difficulty in remembering the muscles of the foot (fig. 4-54) which have a very similar pattern.

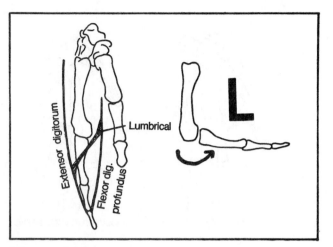

Figure 4-13

Fig. 4-14. The thenar muscles. Although all the hand muscles have been mentioned above, it is of some use to describe the landmark thenar and hypothenar eminences of the hand which contain some of these muscles. In both eminences the letters spell "OAF" as these regions may become quite hypertrophied in the big, hammy hand of a person who is an oaf. These eminences may also atrophy in certain neurological disorders, providing a diagnostic clue as to the presence of a peripheral nerve injury. For example, injury to the median nerve results in a atrophy of the thenar eminence. Injury to the ulnar nerve results in atrophy of the hypothenar eminence.

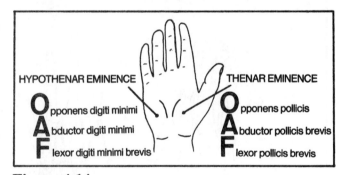

Figure 4-14

B. Muscles That Move the Spine

The muscles that move the spine lie deeper than the muscles that move the scapula and shoulder. The spinal muscles will be considered in two major categories:

Muscles that extend the spine
Muscles that flex the spine

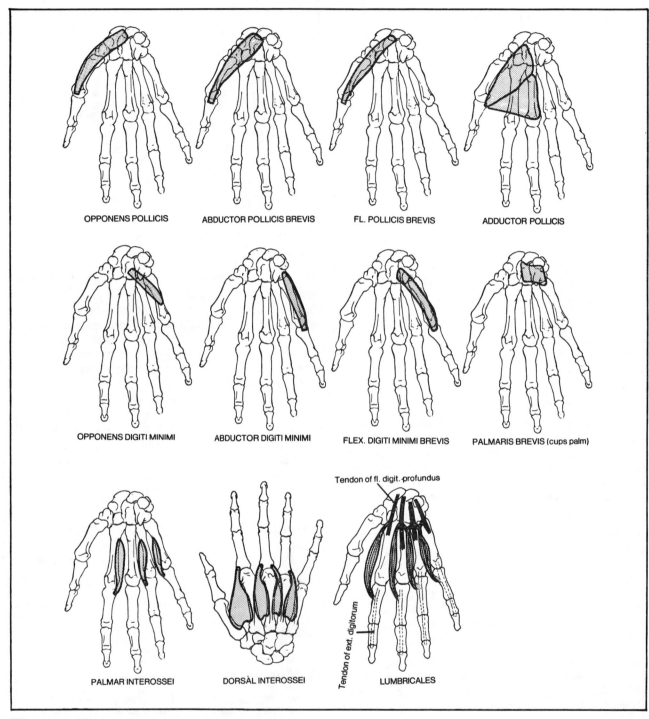

OPPONENS POLLICIS   ABDUCTOR POLLICIS BREVIS   FL. POLLICIS BREVIS   ADDUCTOR POLLICIS

OPPONENS DIGITI MINIMI   ABDUCTOR DIGITI MINIMI   FLEX. DIGITI MINIMI BREVIS   PALMARIS BREVIS (cups palm)

PALMAR INTEROSSEI   DORSAL INTEROSSEI   LUMBRICALES

Tendon of fl. digit. profundus

Tendon of ext. digitorum

Figure 4-12

Muscles that extend the spine

Fig. 4-15. Muscles that extend the spine. The plan of the extensor muscles is logical. Many kinds of connections occur:

(A) interspinous connections - mainly cause extension

(B) intertransverse connections - mainly cause lateral bending when acting unilaterally and extension when acting bilaterally.

(C) and (D) are muscles that connect transverse and spinous processes. These muscles rotate and extend.

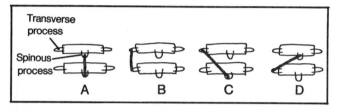

Figure 4-15

Fig. 4-16. Variations in spinal column connections.

(A) Connections may occur between adjacent vertebrae (ab) or non-adjoining vertebrae (ac). Note that muscle connection **ab** is relatively oblique and thus more likely to result in rotation of the spine than is connection **ac**. Connection **ac** is more involved with extension. Line **ab** is shorter than is **ac**. Thus, given the rule of shorter muscles lying deeper, the predominantly rotator muscles of the spine tend to lie very deep.

(B) The rib may be considered an extension of the transverse process and great leverage may be obtained in lateral flexion when muscles connect vertebra to rib.

(C) Muscles may anchor to pelvic bone.

Think of all the possible combinations and you will have a good idea of the general scheme of the hundreds of back muscles. Three main layers of muscles that extend the spine will be considered:

1. Splenius cervicis and capitis: obliquely-running muscles that are the most superficial of the extensor muscles (fig. 4-17).
2. Erector spinae: vertical (erect) muscles that lie intermediate in depth (fig. 4-18).
3. Transversospinalis: obliquely running muscles that connect transverse and spinous vertebral processes. These muscles lie the deepest (fig. 4-19).

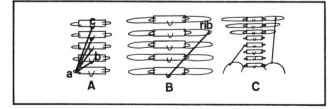

Figure 4-16

Fig. 4-17. The extensor muscles of the spine. Splenius capitis and cervicis. Unlike other spinal muscles, which connect transverse processes with spines of superiorly situated vertebrae, the splenius muscles do the reverse; they are the only ones that connect spinous processes with superiorly situated transverse processes (or lateral skull for the case of the splenius capitis muscle). The splenius muscles are superficial to the erector spinae and transversospinalis muscles. (Who wants them deeper where their peculiar angulation would only disrupt the muscular layout?) The splenius muscles help rotate the head when acting unilaterally. They extend the head and neck, when acting bilaterally.

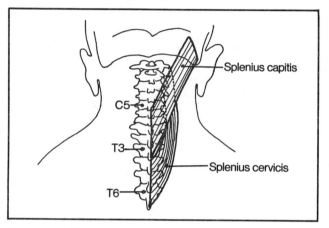

Figure 4-17

Fig. 4-18. The extensor muscles of the spine. The erector spinae muscles (Spinalis, ILiocostalis, LOngissimus). The erector spinae muscles as a whole run relatively parallel to the spine (erect, like a SILO), interconnecting either spinous processes or transverse processes, or extensions of the transverse processes, i.e., ribs and iliac crest).

The **spinalis muscles** mainly interconnect vertebral spines of the **upper 80%** of the vertebral column, including the skull, and predominantly extend the spine. The portion that connects with the skull is an exception in that it originates from transverse processes.

Iliocostalis muscles (costal=ribs) have at least one end attached to a rib and the other end attached to either another rib, a transverse process, or to the sacroilium. These muscles extend the spine (when acting bilaterally) or cause lateral bending (when acting unilaterally). They lie along the **lower 80%** of the spine, including the sacrum and iliac bones.

The **longissimus muscles** predominantly connect transverse processes. When acting unilaterally they cause lateral bending of the spine, but result in extension when contracting bilaterally. Longissimus muscles cover a **long** territory stretching from **sacrum to skull**.

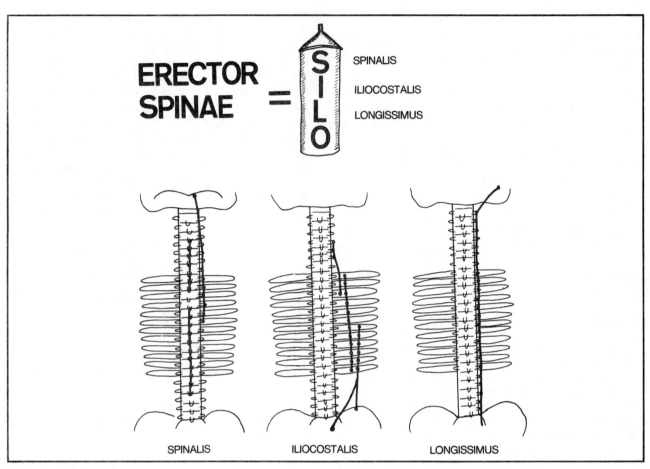

Figure 4-18

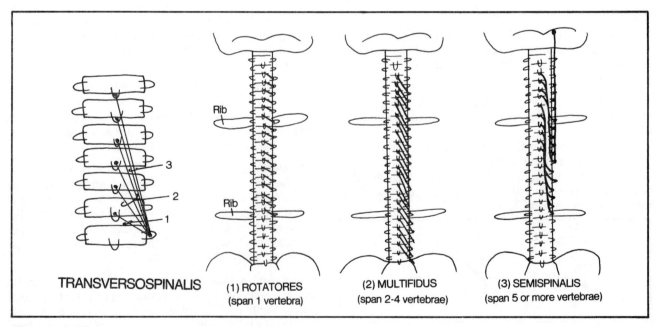

Figure 4-19

Fig. 4-19. The extensor muscles of the spine. Transversospinalis (rotatores, multifidus, semispinalis). These muscles, unlike the straight erector spinae muscles, run relatively obliquely. They connect, in general, a vertebral transverse process with a spinous process above. Therefore, they have a significant rotatory effect. The greatest rotational effect occurs when the muscle is most angulated with the vertical (i.e., when the muscles are shortest). Therefore, these muscles are shorter than the erector spinae muscles and lie more deeply. The shortest of these short muscles (rotatores and multifidus) lie the deepest.

Fig. 4-20. Interspinales and intertransversarii are somewhat difficult to categorize. They are among the deepest lying muscles, but are not oblique. They interconnect adjacent vertebral spines (extension - interspinales) or adjacent transverse processes (extension and/or lateral bending - intertransversarii).

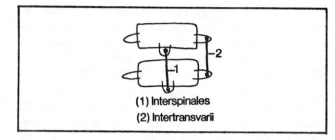

(1) Interspinales
(2) Intertransvarii

Figure 4-20

Muscles that flex the spine

Flexors of the spine are generally located in one of two locations - the neck (fig. 4-21) or the lower back and abdomen (fig. 4-22).

Fig. 4-21. Schematic view of the flexors of the neck. Flexors of the neck fan out like a lampshade. In fact, they

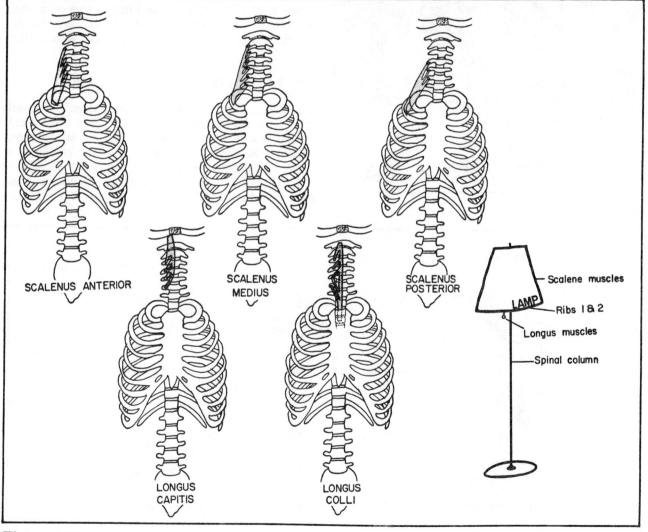

SCALENUS ANTERIOR

SCALENUS MEDIUS

SCALENUS POSTERIOR

LONGUS CAPITIS

LONGUS COLLI

Scalene muscles

Ribs 1 & 2

Longus muscles

Spinal column

Figure 4-21

spell LAMP (L-Longus colli and Longus capitis; A-scalenus Anterior; M-scalius Medius; P-scalenius Posterior). The longus colli and capitis are the string that turns on the light; the scalenus muscles attach the vertebral column to the ribs, and resemble the shade (sort of).

Fig. 4-22. Flexors of the lower back. The PQR muscles: P-Psoas major and minor; Q-Quadratus lumborum; R-Rectus abdominis. The psoas major, in addition, is the strongest flexor muscle of the thigh.

Note that the abdominal muscles also rotate the spine. The combination of right external oblique and left internal oblique muscle contraction exerts a powerful rotatory influence on the spine (see fig. 4-28).

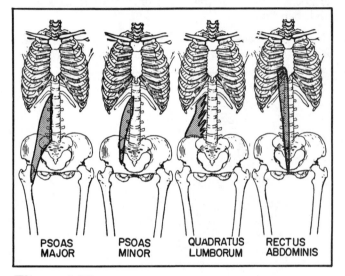

PSOAS MAJOR   PSOAS MINOR   QUADRATUS LUMBORUM   RECTUS ABDOMINIS

Figure 4-22

Little tiny muscles that attach to and move the skull

Some of the extensors of the skull have already been mentioned as they not only move the skull but also extend the cervical spine (splenius capitis, portions of the longissimus, spinalis, and semispinalis muscles). Those which have a more specific function in moving the head, as opposed to the spine, are illustrated in figure 4-23. These muscles directly connect the skull with the atlas or axis. The term "capitis" refers to the "head".

Fig. 4-23. Small muscles that move the skull. The obliquus capitis inferior is the only one of the group that does not attach to the skull. It rotates the head around the dens of the axis. The dotted lines outline the **suboccipital triangle**, which is bounded by obliquus capitis inferior, rectus capitis posterior major, and obliquus capitis superior muscles. The vertebral artery runs in this triangle (see also fig. 6-16). Circulation from vertebral artery to brain may be impeded in this area when there is excessive head turning, in patients with arteriosclerosis (hardening of the arteries). This may result in dizziness.

Don't forget the sternocleidomastoid

Fig. 4-24. The sternocleidomastoid muscle. This muscle is considered separately for 2 reasons. First it is the only muscle that moves the head but does not attach to any vertebra. Second, it is a very important landmark in determining the boundaries of the surgical triangles of the neck (fig. 17-1). If removed surgically, the head can still rotate because other major groupings of muscles can

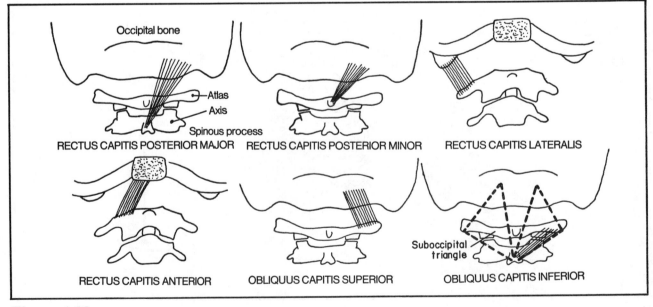

Occipital bone

Atlas
Axis
Spinous process

RECTUS CAPITIS POSTERIOR MAJOR   RECTUS CAPITIS POSTERIOR MINOR   RECTUS CAPITIS LATERALIS

RECTUS CAPITIS ANTERIOR   OBLIQUUS CAPITIS SUPERIOR   OBLIQUUS CAPITIS INFERIOR

Suboccipital triangle

Figure 4-23

influence head rotation (the splenius, erector spinae, sacrospinalis, longus capitis and rectus capitis muscles). Each sternocleidomastoid turns the head to the opposite side. When acting bilaterally, it flexes the head.

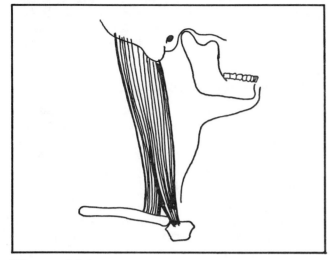

Figure 4-24

Muscles Of Respiration

Muscles of inspiration

During inspiration, air enters the lungs, and the lungs and chest cavity expand. In order to cause air to enter the lungs, a negative pressure must be produced in the chest cavity. **The most important muscle of inspiration is the diaphragm** (fig. 4-25).

The diaphragm is a dome-shaped muscle that attaches to the lower 6 ribs, xiphoid process, and lumbar vertebra to the level of L3. It rises as high as the nipple line. When it contracts it moves downward, decreasing the chest pressure, causing air to enter the lungs.

Fig. 4-25. The diaphragm.

(1) opening for vena cava
(2) opening for esophagus (and vagus nerve trunks); common site of herniation of the stomach (hiatus hernia -fig. 9-5)
(3) opening for aorta (and thoracic duct, and sometimes the azygous vein)
(4) central tendon of diaphragm - fuses with the fibrous pericardium of the heart
(5) opening for psoas muscle
6) vertebrocostal triangle (**foramen of Bochdalek**) - a relatively weak area in the diaphragm which sometimes may be the site of herniation of abdominal viscera
(7) sternocostal hiatus (**foramen of Morgagni**) - the area of penetration of the diaphragm by the superior epigastric vessels (fig. 6-29); rarely the site of herniation of abdominal viscera.

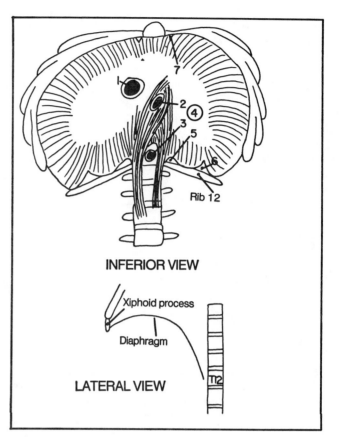

INFERIOR VIEW

Xiphoid process

Diaphragm

LATERAL VIEW

Figure 4-25

The heart lies anteriorly within the thorax and the vena cava enters the right side of the heart. Hence, the vena cava foramen lies relatively anteriorly and to the right. The opening for the aorta lies very posterior, against the vertebral column.

Apart from the diaphragm, other auxiliary muscles contribute to inspiration, by elevating the ribs. Elevating the ribs increases the volume of the chest cavity and creates a negative pressure. The principal is illustrated in figure 4-26.

Fig. 4-26. The bucket and pump handle models of inspiration.

(A) **Elevating the bucket handles** (elevating the lateral aspect of the ribs), increases the distance between the handles (**increases the mediolateral diameter** of the chest cavity).
(B) **Elevation of the hand grip** of the pump handle (the sternal end of the ribs) increases the distance between the hand grip and pump (**increases the anteroposterior diameter** of the chest cavity).

Inspiration involves expansion of the chest cavity through a combination of diaphragmatic contraction and the bucket and pump handle mechanisms.

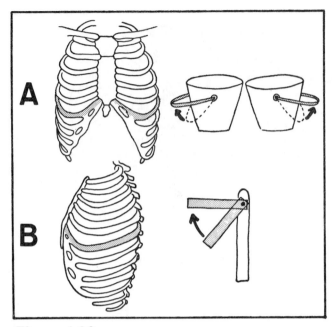

Figure 4-26

Fig. 4-27. Muscles of inspiration that act by elevating the ribs (bucket and pump handle mechanisms). Note that some of the muscles are our old friends that act

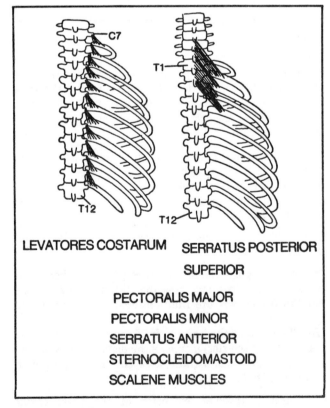

LEVATORES COSTARUM   SERRATUS POSTERIOR

SUPERIOR

PECTORALIS MAJOR
PECTORALIS MINOR
SERRATUS ANTERIOR
STERNOCLEIDOMASTOID
SCALENE MUSCLES

Figure 4-27

primarily to move the upper limb, head, or spine - **pectoralis major** (fig. 4-5) and **minor** (fig. 4-2), **serratus anterior** (fig. 4-2), **sternocleidomastoid** (fig. 4-24), and **scalene** muscles (fig. 4-21). They can also function in an emergency by elevating the ribs. They may be vital to survival in certain chronic pulmonary diseases. Additional muscles not as yet mentioned are the **levatores costarum** and **serratus posterior superior**, shown in figure 4-27.

Muscles of expiration

The expiratory muscles are less important than the inspiratory ones, because **expiration is largely a passive** process. Simply by relaxing, the chest springs back into shape and expiration can occur without any muscle action. Certain muscles do act, however, when forced expiration is attempted (fig. 4-28).

Fig. 4-28. Muscles of expiration. For convenience of illustration, the vertebral column has been split and separated so that an anterior view can show the entire circumference of the chest and abdomen. Note the homology between the muscles of the thorax and those of the abdomen.

(1) external intercostal m. - homologous with the external oblique muscle; most superficial of the intercostal muscles
(2) internal intercostal m. - homologous with the internal oblique; the intermediate muscle layer
(3) transversus thoracis and innermost intercostal m. - homologous with the transversus abdominis; the deepest muscle layer
(4) external oblique m.
(5) internal oblique m.
(6) transversus abdominis m.

The abdominal muscles are the most important muscles of expiration. In addition to them, there are small muscles that connect adjacent ribs (external, internal, and innermost intercostals, and transversus). The functions of the intercostal muscles are somewhat controversial. You would expect them to draw the ribs together. In theory, they could act to induce either inspiration or expiration. For instance, if the first 2 ribs are stabilized by the scalene muscles, the intercostals may act to elevate the ribs. If the lower ribs are stabilized by the quadratus lumborum, the intercostals may depress the ribs.

Some of the muscles that flex or extend the spine (rectus abdominis, quadratus lumborum, and the erector spinae), also can contribute to expiration by depressing the ribs. Not shown is the relatively unimportant serratus posterior inferior muscle. It lies in the same muscle plane as the serratus posterior superior (fig. 4-27). Unlike the latter which elevates the ribs and contributes to inspiration, the serratus posterior inferior muscle angles up from vertebrae T10-L2 to the lower four ribs and acts to depress the ribs.

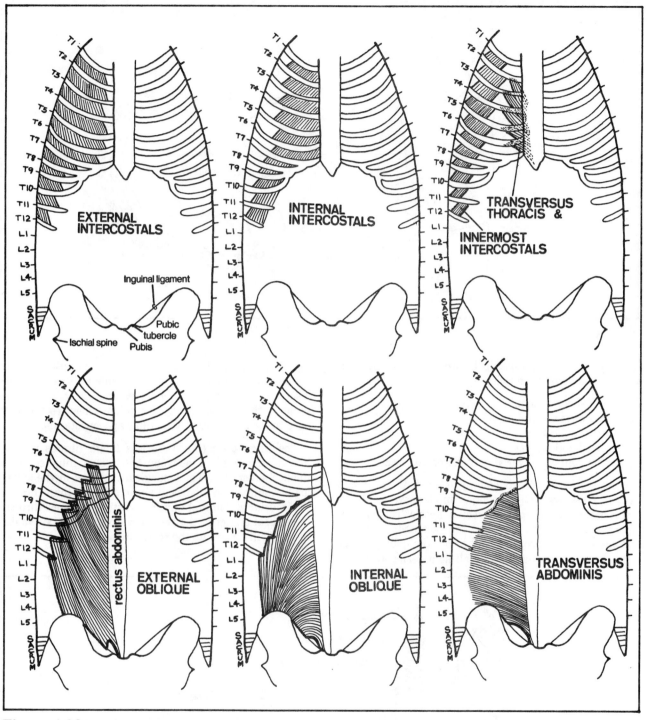

Figure 4-28

Fig. 4-29. The angulation of the external intercostal and external oblique muscles is similar to that of hands in the pocket of a windbreaker jacket. The pockets do not reach the midline, and neither do the external intercostal muscles, for, as figure 4-29 shows, a membrane is positioned between each external intercostal muscle and the sternum. The external oblique muscles don't reach the abdominal midline either as the rectus abdominis blocks the way (see fig. 4-30).

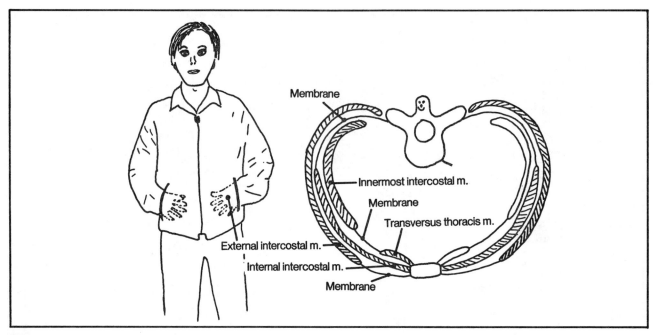

Figure 4-29

Fig. 4-30. The relationship between rectus abdominis and other abdominal muscles.

(1) rectus abdominis m.
(2) transversus abdominis m.
(3) internal oblique m.
(4) external oblique m.
(5) peritoneum with overlying transversalis fascia

A cross section of the abdomen just above the umbilicus (section A) shows that the oblique and transverse abdominal muscles extend up to the rectus muscle. Their fibrous (aponeurotic) sheaths surround the rectus. In the lower abdomen, the fibrous sheaths of the oblique and transversus abdominis muscles all lie anterior to the rec-

tus. This leaves the rectus in direct apposition to the abdominal peritoneum except for some intervening loose fascia. In a sense, the rectus resembles "hands-in-the-pocket", the pocket being the point where rectus goes deep to the transversus abdominis aponeurotic sheath.

The **linea alba** (fig. 4-30) is a tough fibrous vertical line at the midline between the rectus muscles. It is thicker superiorly. Thus, surgical incisions in the superior abdominal midline hold sutures better than those in the inferior abdominal midline. Vertical incisions through the lateral aspect of the rectus muscle should be avoided as they will denervate the rectus muscle.

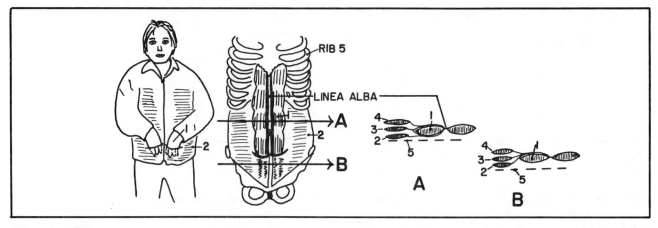

Figure 4-30

Summary Of the Layers Of Back Muscles

Having covered the muscles controlling shoulder movement, movement of the spine, and respiration, certain generalizations can be made. The hundreds of muscles that move the spine are actually quite annoyed that all these other muscles (limb movement, respiration) invade their territory in the back. The back muscles are quite busy. They lie the deepest, near the spine, where they can act. They don't wish to be bothered. They therefore set up a partition, the **thoracolumbar fascia, a** tough sheet of fascia to keep themselves isolated from the more superficial muscles. What have evolved are three main layers of muscles in the back:

1. Superficial muscles (muscles moving the upper limb) - includes muscles attaching the trunk to the scapula, scapula to humerus, and trunk to humerus.
2. Intermediate muscles (of respiration) - the serratus posterior superior and inferior. This is a relatively minor layer.
3. Deep muscles (move the spine). Recall that these deep back muscles are themselves divided into 3 layers:
    a. Splenius muscles (most superficial)
    b. Erector spinae muscles (intermediate) - extension of spine
    c. Transversospinalis muscles (deepest) - extension and rotation of spine.

GOT IT? Not quite. The organizational plan is not quite so neat, as the erector spinae muscles also play some role in respiration and the intercostal respiratory muscles, of course, lie very, very deep.

Fig. 4-31. ADDENDUM: The subclavius muscle. This

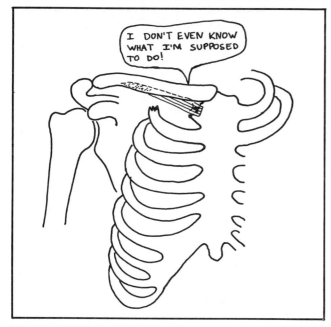

Figure 4-31

relatively minor muscle assists in pulling the clavicle down and forward (maybe).

Muscles Of the Pelvis

Fig. 4-32. The two diaphragms: the thoracoabdominal and pelvic diaphragms. The former contracts in inspiration and separates the contents of thorax and abdomen. The latter contracts in forced expiration, helps constrict the rectum and vagina, and helps prevent the abdominal viscera from falling out the bottom. The pelvic diaphragm separates the pelvis from the perineum. It consists of the levator ani and coccygeus muscles.

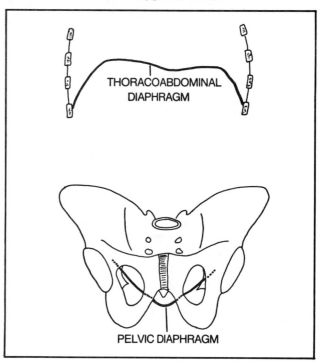

Figure 4-32

Fig. 4-33. Muscles of the pelvic diaphragm (superior view of the the levator ani and coccygeus).

(1) coccygeus m.
(2) levator ani m.
(3) anococcygeal ligament
(4) rectum
(5) central perineal tendon (perineal body)
(6) vagina
(7) urethra

The levator ani attaches to the coccyx, ischial spine, fascia of the obturator internus (figs. 4-45, 4-40), and pubic bone. The fibers of levator ani meet in the midline as a raphe between rectum and coccyx. Some muscle fibers end anterior to the rectum as the perineal body.

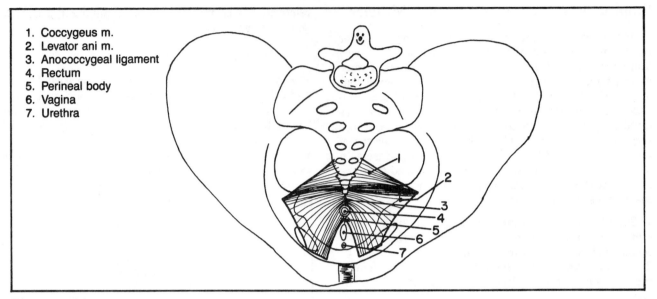

1. Coccygeus m.
2. Levator ani m.
3. Anococcygeal ligament
4. Rectum
5. Perineal body
6. Vagina
7. Urethra

Figure 4-33

Fig. 4-34. Escape from the pelvis. A little imp standing on the pelvic diaphragm has three main escape routes to leave the pelvis:

(A) He can go through the greater sciatic foramen and from there either enter the thigh or curve back under the levator ani muscle to enter the perineum.
(B) He can exit into the thigh by the obturator foramen.
(C) Using the levator ani as a trampoline, he can leap to the top edge of the pubic bone and leave under the inguinal ligament. Important vessels and nerves take these routes.

Muscles Of the Perineum (Doughnuts and a Sandwich)

Figure 4-33 illustrated a view of the pelvic diaphragm from above. The following sequence of figures illustrates it from below, as **the perineum lies inferior to the pelvic diaphragm.**

The perineum may be divided into an **anal** and a **urogenital triangle**.

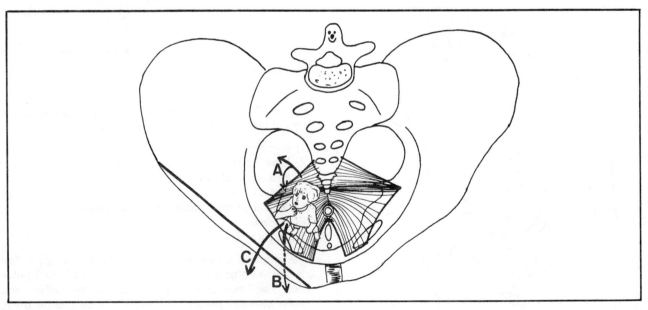

Figure 4-34

Fig. 4-35. The anal and urogenital triangles.

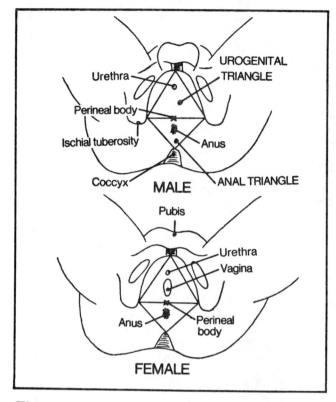

Figure 4-35

Fig. 4-36. The pelvic diaphragm from below (female). The levator ani and coccygeus muscles, which form the pelvic diaphragm, do not form a complete diaphragm; there is a gap around the urethra and vagina. To seal this off we need another diaphragm - the **urogenital diaphragm**, illustrated in figure 4-37. The portion of levator ani muscle that lies closest to the vagina constricts the vagina and may contribute to **vaginismus** (vaginal spasm which may render intercourse difficult).

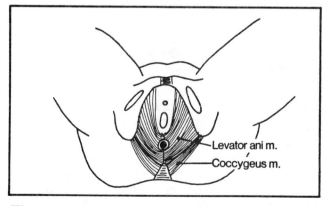

Figure 4-36

Fig. 4-37. The perineum. The urogenital diaphragm is shown laid across the urogenital triangle in the form of a half delicatessen sandwich.

(1) urethra
(2) bulbospongiosus m.
(3) ischiocavernosus m.
(4) superficial transverse perinei m.
(5) inferior (superficial) fascial layer of the urogenital diaphragm (the inferior fascial layer is also called the **perineal membrane**)
(6) deep transverse perinei muscle
(7) superior (deep) fascial layer of the urogenital diaphragm
(8) perineal body - an important fibromuscular structure to which a number of muscles attach (the levator ani, superficial and deep transverse perinei muscles, external anal sphincter and bulbospongiosus muscle)
(9) levator ani m.
(10) external anal sphincter
(11) internal anal sphincter
(12) anococcygeal body (ligament)

The bread of the sandwich (5,7) represents the two fascial layers of the urogenital diaphragm. An olive (bulbospongiosus muscle) (2) rests on the sandwich and a toothpick (urethra) (1) extends all the **way** through the olive and sandwich, back into the pelvic cavity where the prostate (in the male) and bladder are located. Inside the sandwich (not shown) is a round slice of salami (sphincter urethrae muscle) which surrounds the urethra and has apparently been pierced by the toothpick. For the female perineum, stick an extra toothpick into the olive to represent the vagina. Also inside the sandwich is a slice of bacon (deep transverse perinei muscle) (6) which is confined to the cut edge of the sandwich. In the male, two capers (**Cowper's glands = bulbourethral glands**) are also inside the sandwich. These communicate with the urethra. Outside (superficial to) the sandwich is a second slice of bacon (superficial transverse perinei muscle) (4) which also lies near the cut edge of the sandwich. The superficial crust of the bread (3) represents the ischiocavernosus muscle (one on each side).

The olive contains a pimiento - the bulb of the penis (bulb of the vestibule in females). This vascular tissue surrounds the urethra in males and continues along the urethra in males as the corpus spongiosum. In females the bulb surrounds both the vagina and urethra. In males, the bulbospongiosus muscle compresses the bulb, empties the urethra, and secondarily enhances erection, possibly by venous compression. In females the bulbospongiosus muscle constricts the vagina and aids in clitoral erection.

The corpus cavernosum is another body of vascular tissue. It contains two legs, or crura that straddle the corpus spongiosum (fig. 4-38). The ischiocavernosus

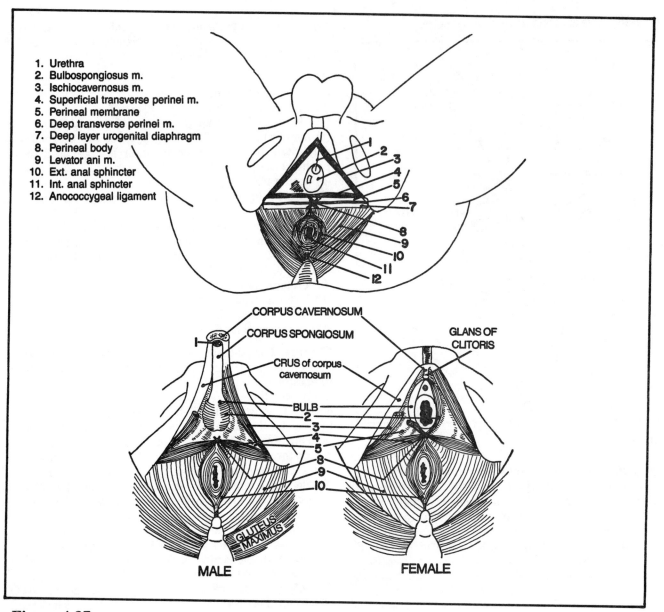

1. Urethra
2. Bulbospongiosus m.
3. Ischiocavernosus m.
4. Superficial transverse perinei m.
5. Perineal membrane
6. Deep transverse perinei m.
7. Deep layer urogenital diaphragm
8. Perineal body
9. Levator ani m.
10. Ext. anal sphincter
11. Int. anal sphincter
12. Anococcygeal ligament

CORPUS CAVERNOSUM
CORPUS SPONGIOSUM
GLANS OF CLITORIS
CRUS of corpus cavernosum
BULB
GLUTEUS MAXIMUS
MALE
FEMALE

Figure 4-37

muscle covers these crura in both male and female, and assists in erection, partly by venous compression.

Within the anal triangle there are several doughnuts, representing the internal(11) and external(10) anal sphincters, which lie within the perineum. The internal sphincter is a doughnut of small diameter, which is composed of involuntary muscle and lies within the anal wall. The external sphincter is a doughnut of larger diameter composed of voluntary muscle that lies outside the anal wall, surrounding the internal sphincter. Part of the external sphincter is oval and extends posteriorly to the coccyx as the anococcygeal ligament and anteriorly to the perineal body.

If the pelvic diaphragm is damaged, it may fail in its supportive role, leading to prolapse of the rectum, bladder and/or uterus.

All skeletal muscles of the perineum connect on one end to bone except for the bulbospongiosus, which does so only during erection (a bad joke).

Fig. 4-38. The erectile areas of the penis. Note the two important erectile tissues - the **corpus spongiosum**, which **surrounds the urethra**, and the **corpus cavernosum** which has two legs (crura) and **lies on top of the corpus spongiosum.**

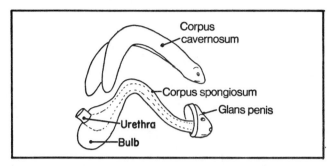

Figure 4-38

Fig. 4-39. Anatomy of the human fly. A human penis seen in cross section about the level of the fly looks like a fly's head with two large cavernous eyes and a spongy mouth (the corpus cavernosum and corpus spongiosum seen in cross section). The corpus cavernosum and spongiosum contain distensile channels that become filled with blood (erection) when the arterial circulation is increased by release of arteriole sphincter action. Enhancement of arterial circulation appears to be more important than compression of venous circulation as a factor in producing erection. The firmness of erection is mainly due to distension of the corpus cavernosum. Distension of the corpus spongiosum is less marked, fortunately, as this might constrict the urethra.

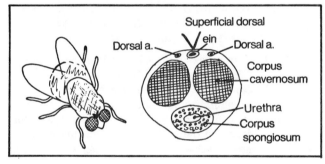

Figure 4-39

Fig. 4-40. Coronal sections through the male and female perineum.

(1) levator ani (pelvic diaphragm)
(2) obturator foramen
(3) obturator internus
(4) superior fascial layer of urogenital diaphragm
(5) urethral sphincter m.
(6) inferior fascial layer (perineal membrane) of urogenital diaphragm
(7) superficial pouch of perineum
(8) anterior extension of ischiorectal fossa (see also fig. 4-41)
(9) crus of penis (clitoris)
(10) ischiocavernosus m.
(11) bulb of penis (bulb of vestibule in females)

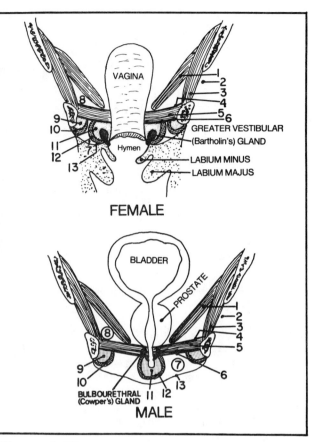

Figure 4-40

(12) bulbospongiogus m.
(13) membranous layer of superficial fascia (Colle's fascia; see fig. 7-5 for description of fascia)

In the male, two capers (Cowper's glands = bulbourethral glands) lie inside the perineal sandwich, embedded in the salami (urethral sphincter m.). These glands connect with the urethra and secret a mucoid component of the seminal fluid. In the female, mucous-secreting **greater vestibular glands (Bartholin's glands)** empty outside the vagina. Sometimes a Bartholin gland duct becomes obstructed, leading to glandular swelling (**Bartholin's cyst**).

Fig. 4-41. Coronal views of rectum and anus.

(1) obturator internus m.
(2) levator ani m.
(3) puborectalis portion of levator ani m. - felt as a sharp ridge on rectal exam
(4) longitudinal muscle of rectum and anal canal
(5) circular muscle of rectum
(6) internal (involuntary) sphincter of anal canal
(7) external (voluntary) sphincter of anal canal (deep, superficial and subcutaneous portions, respectively)
(8) ischiorectal fossa - the fat-filled space between the

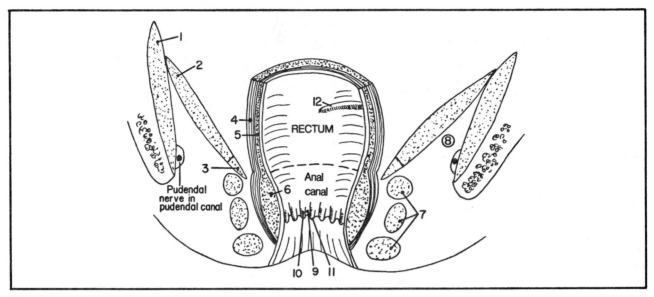

Figure 4-41

levator ani muscle and skin. It is most predominant in the anal triangle. The ischiorectal fossa is easily compressed, allowing space for expansion of an anus filled with feces, or the vagina during childbirth.

(9) anal sinuses (crypts) - recesses lying between the anal columns. These crypts contain lubricating mucous glands that may become inflamed ("cryptitis")

(10) anal columns - elevated folds

(11) anal valves - small folds at the distal end of the anal sinuses. The line of anal valves is referred to as the **pectinate line** because (not really) if one "pecks" below it, it hurts, but it doesn't hurt if one "pecks" above it. I.e., the area below the pectinate line is very sensitive to pain, but the area above it is not. Thus, **external hemorrhoids**, which **lie below the pectinate line**, frequently hurt or itch, whereas **internal hemorrhoids**, which **lie above the pectinate line**, do not. The autonomic nervous system innervates the area above the pectinate line (hence, minimal pain sensation), whereas the somatic nervous system innervates below.

(12) lower transverse rectal fold. In ascending through the rectum, as with a sigmoidoscope, three folds in the wall of the rectum are encountered, a lower transverse fold (illustrated) on the right, a middle fold on the left, and then an upper transverse fold on the right. The folds are convenient sites for rectal biopsy, as there is decreased risk of rectal perforation.

The **rectum** lies totally within the pelvis. It becomes the **anal canal** by passing the levator ani muscle. The **anus** is the terminal opening of the anal canal.

During childbirth, the obstetrician may perform an **episiotomy** procedure, in which an incision is made in the posterior vaginal wall (fig. 4-37) to enlarge the vagi-

nal orifice. This prevents irregular tears that might damage the rectum or external anal sphincter. The incision may be carried out directly posteriorly (through the perineal body) from the posterior aspect of the vaginal orifice, stopping short of the external anal sphincter. Alternatively, in order to insure that the external anal sphincter is not damaged by the incision, the episiotomy may be extended posterolaterally into the ischiorectal fossa, taking care not to damage the levator ani.

Fig. 4-42. The puborectalis sling. The anterior pull of the puborectalis portion of the levator ani (also see fig. 4-41) is responsible for the angulation between anus and rectum. This angulation is important to know about in performing a sigmoidoscopy. The sigmoidoscope should be angulated toward the umbilicus on entering the anus, and then angled backward, when passing into the rectum.

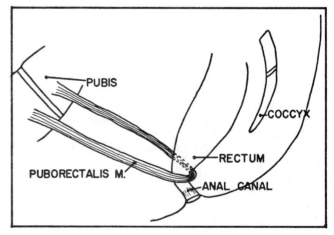

Figure 4-42

The puborectalis, by producing the angle between rectum and anus, helps prevent stool from applying excessive pressure on the anal canal. It relaxes during defecation.

Fig. 4-43. The cremaster muscle, an extension of the internal oblique muscle. It attaches to the spermatic cord and elevates the testes. The "cremaster reflex" is a normal elevation of the testes on scratching the upper medial thigh.

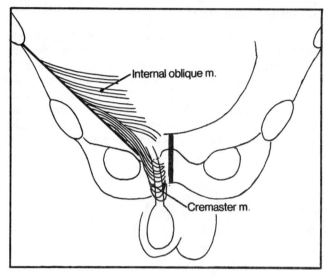

Figure 4-43

Muscles Of the Lower Extremity

ABDUCTORS OF THE HIP
obturator internus (gemellus inferior and superior)
gluteus minimus
gluteus medius
tensor fascia latae
piriformis

ADDUCTORS OF THE HIP
gluteus maximus
obturator externus
quadratus femoris
pectineus
adductors brevis, longus, and magnus
gracilis

FLEXORS OF THE HIP
psoas major
iliacus
sartorius
rectus femoris

EXTENSORS OF THE HIP
gluteus maximus
biceps femoris
semitendinosus
semimembranosus

FLEXORS OF THE KNEE
gracilis
sartorius
biceps femoris
semitendinosus
semimembranosus

EXTENSORS OF THE KNEE
vastus medialis
vastus intermedius
vastus lateralis
rectus femoris

Muscles that move the hip and knee are listed above according to function. A number of these muscles act at more than one joint and have more than one function. Consequently they are easier to remember by region, than by function, namely:

1. Iliac muscles
2. Gluteal muscles

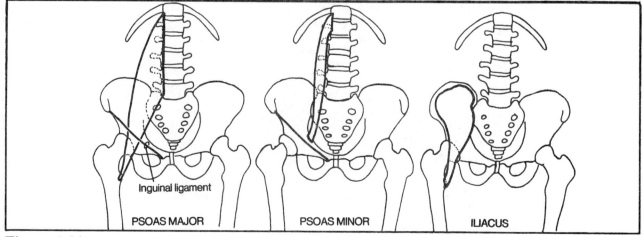

Figure 4-44

3. Muscles of the medial thigh
4. Muscles of the anterior thigh
5. Muscles of the posterior thigh.

Note the absence of a grouping entitled "Muscles of the lateral thigh"; there simply isn't enough room for this group. If you learn these five categories you will know the main activators of the hip and knee joints. Study figures 4-44 through 4-50. .

Fig. 4-44. The iliac muscles. Note that although there are 3 iliac muscles, the psoas minor does not act on the lower extremity. It, like the psoas major, contributes to flexion of the spine. It just doesn't belong in our important discussion of the lower extremity. In fact, the psoas minor is absent about 40% of the time. The psoas major and iliacus muscle together are sometimes referred to as the iliopsoas muscle.

(1) psoas major - flexes the thigh and vertebral column. The appendix overlies the psoas major. Thus, in appen-dicitis, the patient often experiences pain on hyperextension of the thigh, a useful sign on examination. In marked trauma, stretching of the psoas major muscle may acutely avulse it from its attachment to the lesser trochanter. The psoas major normally protrudes through the diaphragm into the thorax (fig. 4-25). Thus, thoracic infections (such as pulmonary tuberculosis) sometimes spread along the psoas major and may present as enlarged lymph nodes in the inguinal region.

(2) psoas minor - flexes the vertebral column
(3) iliacus - flexes the thigh

Fig. 4-45. The gluteal muscles (muscles of the rump). Note the broad outline of the gluteus maximus, the largest and most superficial of the gluteal muscles. All of the gluteal muscles, except for the gluteus maximus, point roughly in the direction of, or connect outright with, the greater trochanter of the femur. I.e., the **insertions are more or less in the same place, whereas the origins travel quite a bit.** Play follow the dots; by connecting

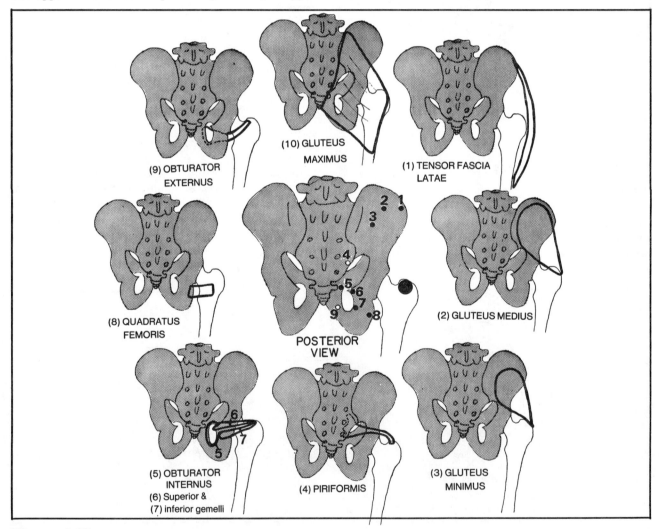

(9) OBTURATOR EXTERNUS

(10) GLUTEUS MAXIMUS

(1) TENSOR FASCIA LATAE

(8) QUADRATUS FEMORIS

POSTERIOR VIEW

(2) GLUTEUS MEDIUS

(5) OBTURATOR INTERNUS
(6) Superior &
(7) inferior gemelli

(4) PIRIFORMIS

(3) GLUTEUS MINIMUS

Figure 4-45

numbers 1-9 you will follow the course of the origins of the gluteal muscles. In progressing from 1-9, the muscles begin as abductors and end up as adductors. They also act as rotators. Gluteus maximus adducts, extends, and laterally rotates the thigh; it is used in **forced** extension, as in arising from a chair, rather than in normal walking (test this on yourself). The hamstrings are more important extensors for ordinary walking.

Fig. 4-46. Sitting. When one sits, one sits neither on the coccyx nor on the gluteus maximus. One sits on the **ischial tuberosity.**

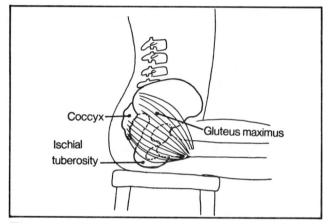

Figure 4-46

Fig. 4-47. The medial thigh muscles. These adduct and flex the thigh. The origins of the gluteal muscles roam about, but the **origins of the medial thigh muscles tend to remain stationary, around the obturator foramen, whereas their insertions travel down the femur.** They travel down so far that finally one muscle, gracilis, arrives at the leg. This introduces us to a new type of lower extremity muscle - one that spans two joints. A number of the anterior and posterior thigh muscles are of that type.

The term "pulled groin" refers to the stretching or tearing of the adductor muscles at their tendinous origins around the obturator foramen.

Fig. 4-48. Muscles of the anterior thigh. Continuing from the medial group of muscles, which ended with gracilis, the sartorius muscle marks the beginning of the anterior group, which is presented in a medial to lateral direction.

(1) sartorius (the tailor's muscle) - flexes the thigh and knee, simulating the position of a tailor sitting cross-legged at work
(2) vastus medialis
(3) vastus intermedius
(4) rectus femoris - overlies the vastus intermedius as expected, since the rectus is longer
(5) vastus lateralis

The vastus muscles span only the knee joint and contribute to knee extension. The **sartorius** and **rectus femoris span both hip and knee**; although they both flex the thigh, the rectus femoris **extends** the knee whereas the sartorius **flexes** the knee.

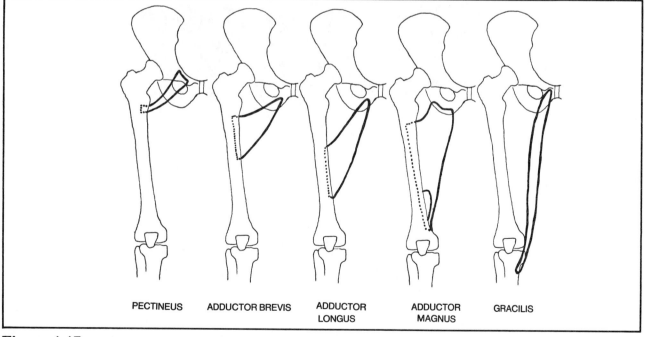

PECTINEUS   ADDUCTOR BREVIS   ADDUCTOR LONGUS   ADDUCTOR MAGNUS   GRACILIS

Figure 4-47

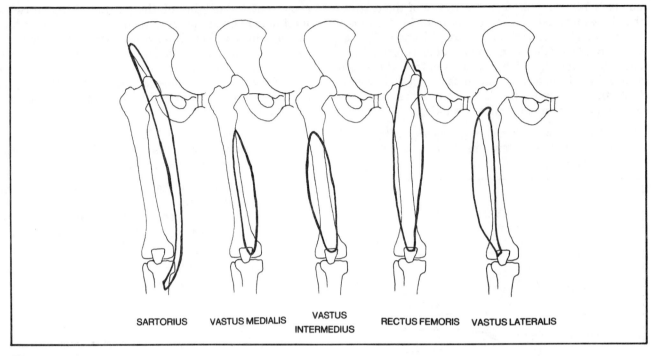

SARTORIUS   VASTUS MEDIALIS   VASTUS INTERMEDIUS   RECTUS FEMORIS   VASTUS LATERALIS

Figure 4-48

Fig. 4-49. Muscles of the posterior thigh (hamstrings). All of these muscles span both hip and knee. Imagine a dotted line extending along the midline of the femur. The BIceps muscle ("BI=2) occupies BOTH sides of the line, crossing the midline from medial to lateral to reach the fibula and lateral condyle of the tibia. The "SEMI" muscles ("SEMI=partial) only occupy part of this territory, remaining only on the medial side of the line.

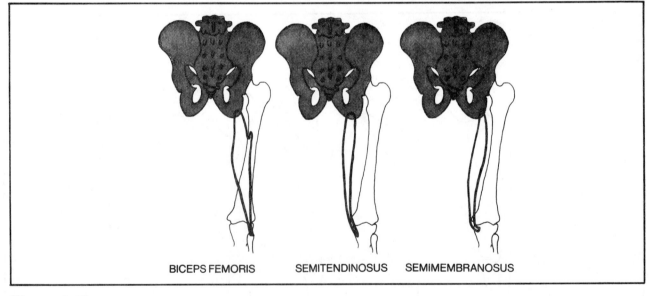

BICEPS FEMORIS   SEMITENDINOSUS   SEMIMEMBRANOSUS

Figure 4-49

Fig. 4-50. The popliteus muscle. This muscle doesn't quite fit into any of the categories that were discussed and therefore is presented separately. It helps to flex ("unlock") the knee.

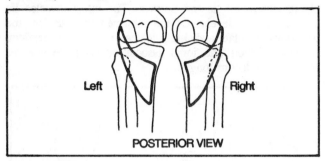

Figure 4-50

Muscles Of the Leg and Foot

Muscles of the leg and foot are presented in the following categories:

1. Leg muscles that plantar flex the ankle.
2. Leg muscles that primarily invert or evert the ankle (and also do some ankle flexion or extension).
3. Leg muscles that flex or extend the toes.
4. Intrinsic muscles of the foot

Fig. 4-51. Muscles that primarily plantar flex the ankle (e.g. standing on tiptoe). These are presented as a special category as they are quite strong (most people can lift several hundred pounds by ankle plantar flexion).

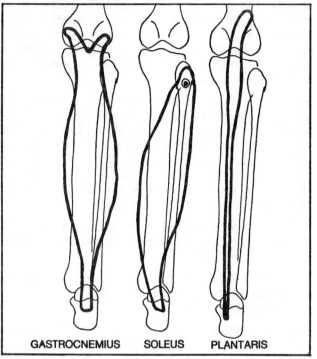

Figure 4-51

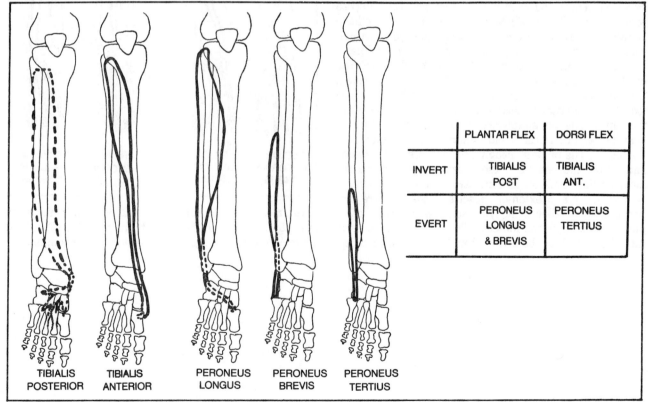

	PLANTAR FLEX	DORSI FLEX
INVERT	TIBIALIS POST	TIBIALIS ANT.
EVERT	PERONEUS LONGUS & BREVIS	PERONEUS TERTIUS

Figure 4-52

The main muscles of plantar flexion, the **gastrocnemius** and **soleus**, are quite prominent. The gastrocnemius lies superficial to the soleus as it is longer. These muscles are relatively large and long when compared with those of the upper extremity. The soleus looks like a sole, or flatfish. The gastrocnemius and soleus muscles share a common tendon (tendo calcaneus) which not uncommonly tears (Achilles' tendon).

Fig. 4-52. Muscles that primarily invert or evert the ankle (i.e., invert or evert the bottom of the foot). These also function in flexion and extension of the ankle. The pEroneus muscles **Evert** the ankles, thereby moving the knees together as a "PAIR-O-KNEES". The tIbialis muscles Invert that ankles. The function of dorsiflexion or plantar flexion depends on whether or not the tendon passes anterior or posterior to a malleolus.

The peroneus longus and brevis tendons pathologically may slip anterior to the lower end of the fibula. This is surgically correctable.

Fig. 4-53. Muscles of the leg that primarily flex or extend the toes. These also assist in ankle plantarflexion and dorsiflexion.

Fig. 4-54. Muscles intrinsic to the foot. The intrinsic muscles of the foot resemble very closely the muscles of the hand (compare with fig. 4-12) with the exception that **the foot contains no opponens** (pollicis or digiti minimi) **muscles.** (Have you ever noticed that you can't oppose your big and little toes?) Nor does the foot need or have a **palmaris brevis** muscle. (Do you think you can cup your foot like you can cup your palm with the palmaris brevis muscle?) **These muscles are traded off for intrinsic digital extensor and flexors**, as follows. Digital extensors and flexors originate in both the leg and foot, but no digital extensors and flexors originate from the femur. The toes simply aren't used enough to warrant long, powerful extensors and flexors that originate from the femur. The fingers, though, require excellent extensor and flexor movements; these muscles originate as far away as the humerus. In contrast to the upper extremity, the flexor and extensor muscles of the toes are proportionately smaller, some lying solely within the foot, in place of the opponens muscles. The intrinsic muscles in the sole of foot may be divided into four layers. In gener-

al, the longer muscles are more superficial. Thus, **the most superficial grouping contains flexor digitorum brevis, abductor hallucis and abductor digiti minimi,** which stretch from the calcaneus to the phalanges. The plantar interossei lie the deepest. The other muscles lie at intermediate levels. Of the intermediate muscles, the quadratus plantae and lumbricales lie more superficial because of their additional association with the tendons of flexor digitorum longus.

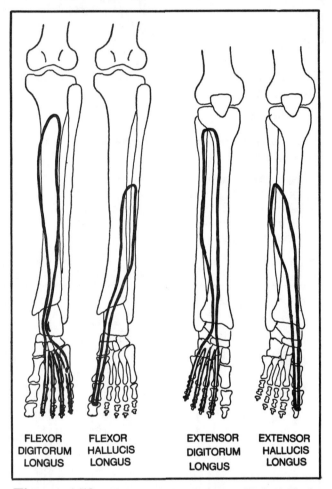

FLEXOR DIGITORUM LONGUS

FLEXOR HALLUCIS LONGUS

EXTENSOR DIGITORUM LONGUS

EXTENSOR HALLUCIS LONGUS

Figure 4-53. Passing behind the medial malleolus are "Tom, Dick, and Harry": tibialis posterior (Fig. 4-52), flexor digitorum longus, and flexor hallucis longus.

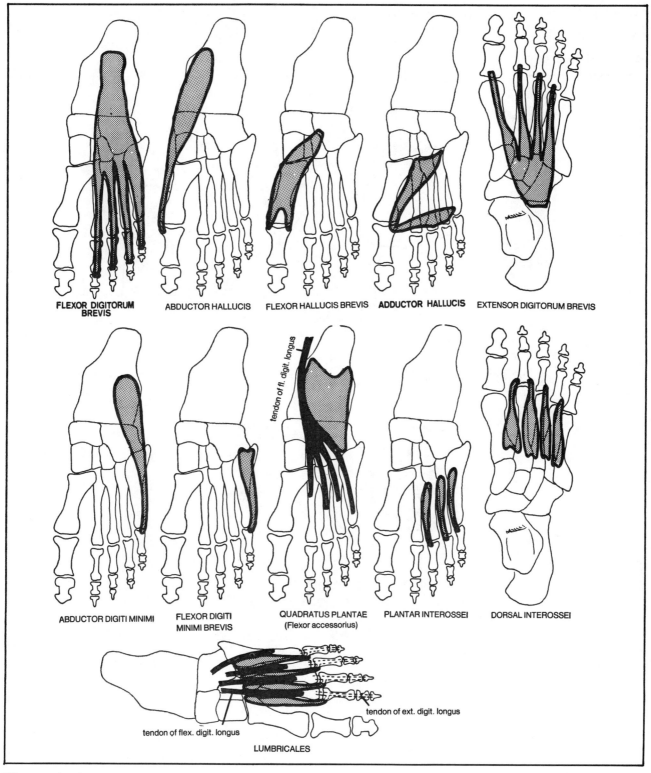

Figure 4-54

53

Muscles Of the Head and Neck (see also figs. 17-1 through 17-4)

Muscles of the face

Fig. 4-55.   Muscles of the upper face.

Fig. 4-56.   Muscles of the nose.

Fig. 4-57.   Muscles of the lips.

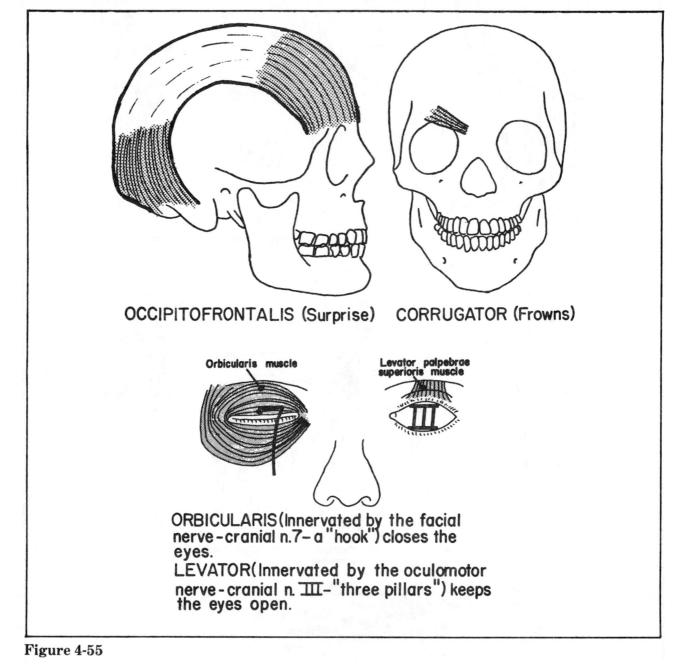

OCCIPITOFRONTALIS (Surprise)   CORRUGATOR (Frowns)

Orbicularis muscle

Levator palpebrae superioris muscle

ORBICULARIS(Innervated by the facial nerve–cranial n.7– a "hook") closes the eyes.
LEVATOR(Innervated by the oculomotor nerve–cranial n. III–"three pillars") keeps the eyes open.

Figure 4-55

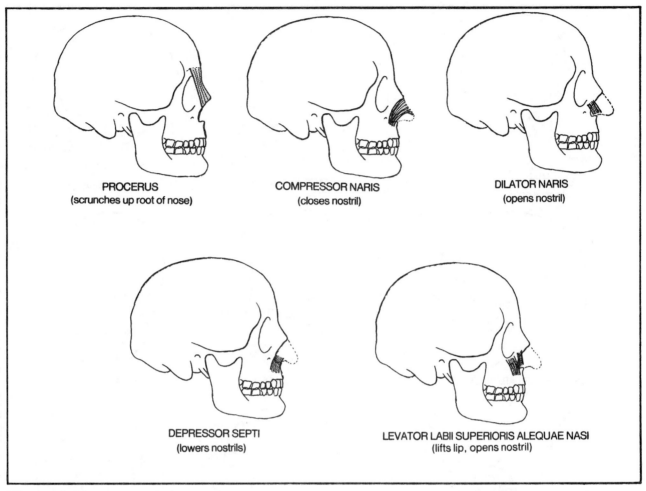

PROCERUS
(scrunches up root of nose)

COMPRESSOR NARIS
(closes nostril)

DILATOR NARIS
(opens nostril)

DEPRESSOR SEPTI
(lowers nostrils)

LEVATOR LABII SUPERIORIS ALEQUAE NASI
(lifts lip, opens nostril)

Figure 4-56

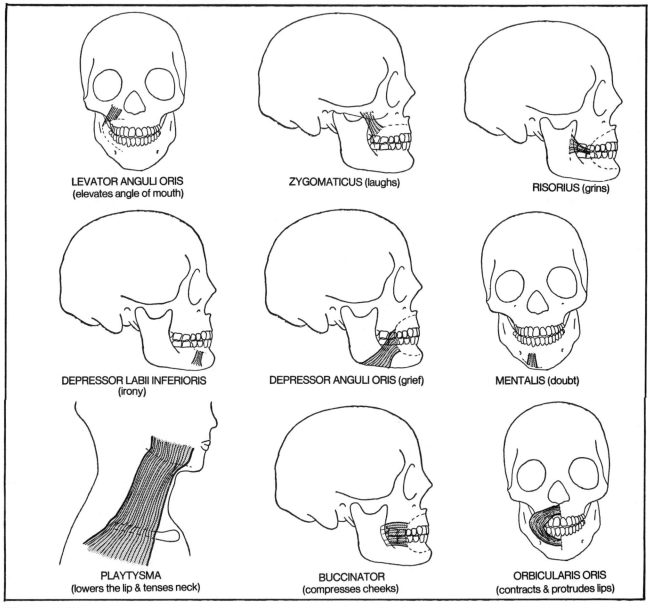

LEVATOR ANGULI ORIS
(elevates angle of mouth)

ZYGOMATICUS (laughs)

RISORIUS (grins)

DEPRESSOR LABII INFERIORIS
(irony)

DEPRESSOR ANGULI ORIS (grief)

MENTALIS (doubt)

PLAYTYSMA
(lowers the lip & tenses neck)

BUCCINATOR
(compresses cheeks)

ORBICULARIS ORIS
(contracts & protrudes lips)

Figure 4-57

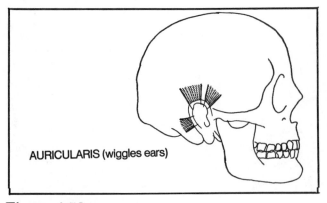

AURICULARIS (wiggles ears)

Figure 4-58

Fig. 4-58. Muscles moving the ears.

Fig. 4-59. Chewing muscles. The pterygoid muscles, when acting unilaterally, move the jaw to one side.

The facial muscles are all innervated by the facial nerve (cranial nerve 7), except for the chewing muscles, which are innervated by the trigeminal nerve (cranial nerve 5), and the levator palpebrae superioris muscle (oculomotor nerve - cranial nerve 3) which elevates the eyelid (see also fig. 15-1).

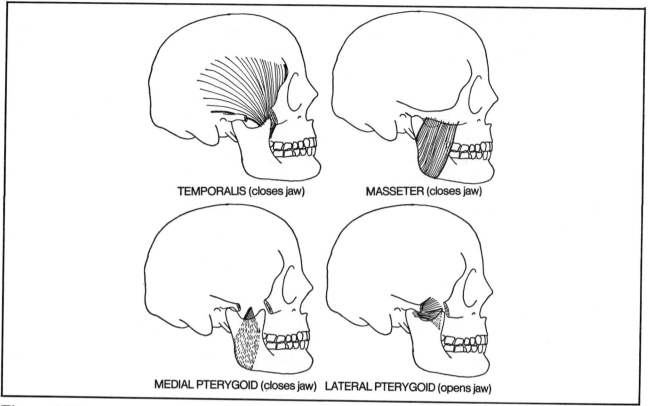

TEMPORALIS (closes jaw)   MASSETER (closes jaw)

MEDIAL PTERYGOID (closes jaw)   LATERAL PTERYGOID (opens jaw)

Figure 4-59

Fig. 4-60. Muscles of the eye. S.O., superior oblique; I.O., inferior oblique, S.R., superior rectus; I.R., inferior rectus; M.R., medial rectus; L.R., lateral rectus. Each extrinsic muscle of the eye attaches to the eye as well as to the nasal aspect of the orbit. None of the eye muscles attach to the temporal wall of the orbit. All the muscles, including the levator palpebrae muscle, originate in the tendinous ring on the bone that surrounds the optic foramen, with the exception of the inferior oblique, which originates at the nasal aspect of the maxillary bone (fig. 2-29).

The mnemonic for innervation of the eye muscles is "LR6, SO4" (Lateral Rectus - cranial nerve 6, Superior Oblique - cranial nerve 4). The other muscles, including the levator palpebrae superioris, are innervated by cranial nerve 3.

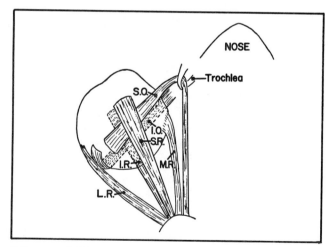

Figure 4-60

Fig. 4-61. Table of eye muscle actions.

Eye Muscle	Nerve	Primary Function	Deficit
Medial rectus	Oculomotor (CN3)	Moves eye nasally	Eye is down and out because of unopposed action of lateral rectus and superior oblique
Lateral rectus	Abducens (CN6)	Moves eye temporally	Eye cannot look temporally
Superior rectus	Oculomotor (CN3)	Moves eye up	Weakness of upward gaze
Inferior rectus	Oculomotor (CN3)	Moves eye down	Weakness of downward gaze
Superior oblique	Trochlear (CN4)	1) Moves eye down when eye is already looking nasally. 2) Rotates eye when eye is already looking temporally. 3) Moves eye down and out when eye is in straight ahead position.	Vertical diplopia Head tilt (compensation for imbalance of rotation).
Inferior oblique	Oculomotor (CN3)	1) Moves eye up when eye is already looking nasally. 2) Rotates eye when eye is already looking temporally. 3) Moves eye up and out when eye is in straight ahead position.	Vertical diplopia Head tilt
Levator palpebrae superioris	Oculomotor (CN3)	Elevates upper lid	Marked ptosis
Muller's muscle (see Fig. 15-1)	Cervical sympathetics	Elevates upper lid	Mild ptosis

Figure 4-61

Fig. 4-62. Action of the eyelid muscles. The orbicularis muscle (innervated by the facial nerve - cranial nerve 7, a hook) closes the eyes. The levator muscle (innervated by the oculomotor nerve - cranial nerve III, three pillars) keeps the eyes open. If you seem to recall seeing this picture in figure 4-55, you are right. This point has been repeated here because it is clinically very important to remember which muscles open and close the eye. Patients not infrequently present with difficulties in these activities.

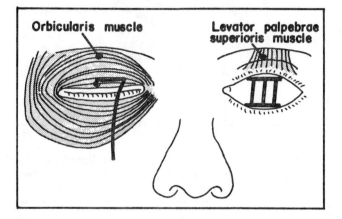

Figure 4-62

Fig. 4-63. Muscles of hearing, as seen from the middle ear cavity (fig. 16-3).

(1) tensor tympani - innervated by cranial nerve 5
(2) stapedius - innervated by cranial nerve 7

There are two muscles of hearing. Each hearing muscle attaches at one end to a fixed bone and at the other end to one of the small bones of hearing, either the malleus or stapes. These muscles help stabilize the malleus and stapes and prevent sound waves from being heard too loudly. Poor incus, the middle bone, has no muscle. The chorda tympani nerve, a branch of the facial nerve (cranial nerve 7) mediates taste. It runs in the crotch formed by the junction of malleus and incus.

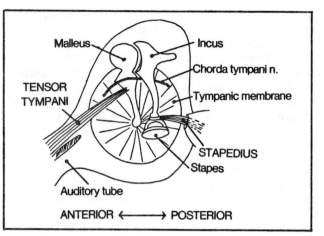

Figure 4-63

Muscles of speech

1. ALTER SPACE BETWEEN VOCAL CORDS:
   posterior cricoarytenoid - opens glottis
   lateral cricoarytenoid - closes glottis
2. ALTER TENSION OF VOCAL CORDS:
   Cricothyroid - elongates and tightens vocal cords
   Thyroarytenoid and vocalis - shorten vocal cords
3. PROTECT ENTRANCE TO LARYNX:
   Thyroepiglotticus
   Aryepiglottic muscles

This section deals predominantly with the muscles of the larynx although it should be noted that other muscle groupings also affect the quality of speech. E.g., the "KLM" sounds: the sounds "Kuh, Kuh, Kuh" test soft palate elevation (levator palati - cranial nerve 10); "La, La, La" tests the tongue (cranial nerve 12) and "Mi, Mi, Mi" tests the lips (cranial nerve 7). In addition, the expiratory muscles help control loudness. The laryngeal muscles control **pitch**.

The ingredients necessary to understand the structure of the larynx and trachea include: an assortment of rings, a pair of boots, a facemask and a shoehorn (fig. 4-64).

Fig. 4-64. Components of the larynx and trachea.

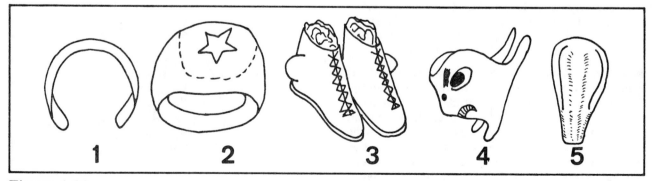

Figure 4-64

(1) tracheal rings - open posteriorly; as the esophagus lies posterior to the trachea, esophageal tumors may easily invade the larynx
(2) cricoid cartilage (a signet ring - closed all around)
(3) arytenoid cartilages (a pair of boots). In these particular boots the lateral sides are bulged out, as if the wearer had very prominent lateral malleoli. Protruding out through each shoe is a sock (**corniculate cartilage**)
(4) thyroid cartilage (a face mask) - open posteriorly
(5) epiglottis (a shoehorn)

Fig. 4-65. Organization of the laryngeal and tracheal cartilages. The signet ring (cricoid cartilage) sits above the highest tracheal ring. Its wide (posterior) portion provides a platform on which the boots (arytenoid cartilages) rest. The boots point anteriorly, toward the shoehorn (epiglottis). The face mask completely hides the boots but hides only part of the signet ring. The shoehorn protrudes above the mask, even above the top of the overlying hyoid bone.

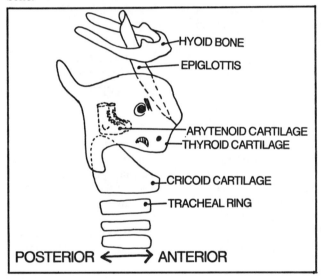

Figure 4-65

Fig. 4-66. Ligaments of the larynx. Most ligaments (L) are named according to the structures they interconnect (including such names as the hyoepiglottic, thyrohyoid, thyroepiglottic, cricoarytenoid, lateral and median cricothyroid, and cricotracheal ligaments). The vocal ligament is shown as a line connecting the front tip of the boot (arytenoid cartilage) with the thyroid cartilage.

In an **emergency** tracheostomy, a wide-bore needle is passed into the trachea through the **median cricothyroid ligament**. Under more controlled circumstances, the procedure involves an incision through the second and third tracheal cartilages, in order to avoid damage to the laryngeal structures. The may require an incision through the **isthmus** (midline) of the thyroid gland, which lies at tracheal rings 2 and 3. It is better, however, to cut through the isthmus than to risk hemorrhage by cutting into the rich venous plexus that lies below the thyroid gland.

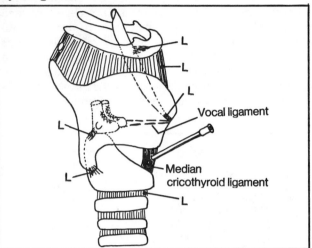

Figure 4-66

Fig. 4-67. Superior view of the vocal ligament. (A) shows associated cartilages. (B) shows, in addition, the conus elasticus and median cricothyroid ligament. The

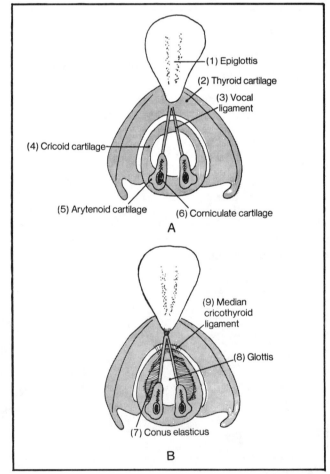

Figure 4-67

conus elasticus is a fibroelastic membrane connecting the vocal ligament and cricoid cartilage. It is actually shaped more like a pup tent than a cone. The tent's roof closes when the **glottis** (the space between the vocal ligaments) closes.

Fig. 4-68. The aryepiglottic fold - a fold stretching between arytenoid cartilage and epiglottis (compare with fig. 4-69). It may be pictured as follows:

A. Imagine a barrel with its lid open and a pair of boots resting on the edge of the barrel.

B. Imagine someone trying to stuff a long garbage bag down into the barrel.

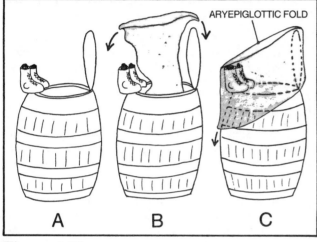

Figure 4-68

C. Imagine the top edge of the bag being doubled over to fold around the barrel. The boots are the arytenoid cartilage. The rings of the barrel are the cricoid and tracheal rings. The upper rim of the barrel is the vocal cords. The lid is the epiglottis. The inner lining of the bag is the mucosal surface, which after the bag is folded over, is present both inside and outside the barrel. The mucosa inside the barrel is laryngeal mucosa. The mucosa outside the barrel is pharyngeal mucosa. The fold between the boots and epiglottis is the aryepiglottic fold which is actually a double layer of mucous membrane. Sandwiched between the two layers of folded-over bag are the cartilages, ligaments and muscles of the larynx.

Fig. 4-69. A. The laryngeal cartilages, as seen from behind.

B. Posterior view of the larynx covered by mucosa, showing the aryepiglottic folds. The posterior wall of the pharynx has been removed. The region of pharynx that lies at the level of the larynx is called the **laryngopharynx** or **hypopharynx**. Progressively higher levels are termed **oral pharynx** and **nasopharynx**.

Fig. 4-70. The vestibular ligament ("false vocal cord"). A further refinement in the description of laryngeal anatomy is the presence of two vocal ligaments on each side - a "true" vocal ligament and a "false", or vestibular ligament. The true vocal ligament connects the front tips of the boots with the thyroid cartilage. The vestibular ligament connects the laces of the boots with the thyroid cartilage, as in A. The mucosa lining the inside of the

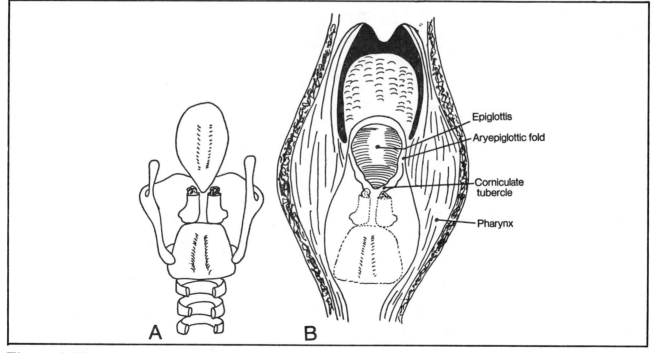

Figure 4-69

larynx forms a pocket or ventricle between the two ligaments (as shown in coronal section B). Thus, when the physician looks down into the larynx using a mirror, inset C is seen. There is a **true vocal cord** (associated with the **vocal ligament**) and a **false cord**, associated with the **vestibular ligament**. It is the true cord that is associated with pitch regulation in speech.

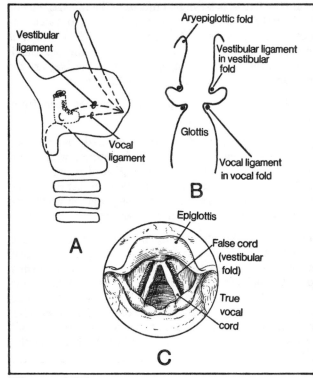

Figure 4-70

Fig. 4-71. Opening of the glottis by the posterior cricoarytenoid muscle. The glottis opens in two ways:

1. The boots (arytenoid cartilages) move apart (imagine a person adopting a wide-based stance). The **vertical** fibers of the posterior cricoarytenoid muscle perform this function.
2. The boots rotate so that the anterior tips of the boots, but not the heels, move apart. The **horizontal** fibers of the posterior cricoarytenoid muscles do this.

The **posterior cricoarytenoid muscles are the most important skeletal muscles in the body. They are the only muscles that open the glottis.** If they become paralyzed, the glottis remains closed and air cannot enter the lungs.

Fig. 4-72. Muscles closing the glottis. The glottis is closed partly by the **lateral cricoarytenoid muscle**, which has a rotatory action, enabling the anterior tips of the boots to click together. In addition, the **oblique and transverse arytenoid muscles** move the boots as a whole together.

Fig. 4-73. Muscles that control the tension of the vocal ligament (lateral view). Such muscles move the thyroid cartilage either away from or toward the arytenoid cartilage.

The **thyroarytenoid** and **vocalis** muscles decrease the distance between thyroid cartilage and arytenoid cartilage and alter the tension of the vocal cord. The thyroarytenoid, in addition, **closes** the glottis by rotating the anterior tips of the boots together.

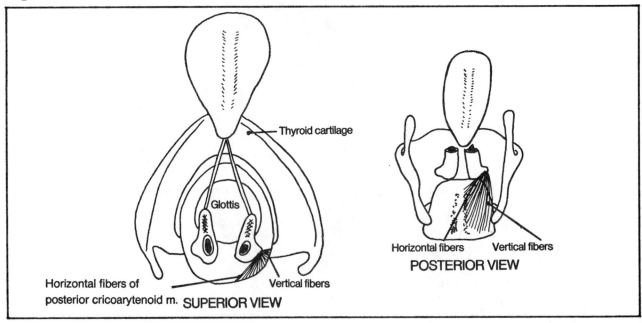

Figure 4-71

The **cricothyroid muscle** increases the distance between thyroid cartilage and arytenoid cartilage, and increases the tension of the vocal cords.

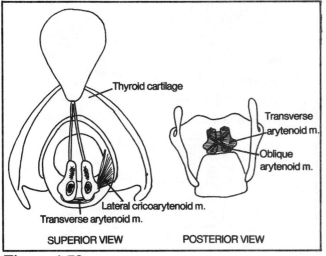

Figure 4-72

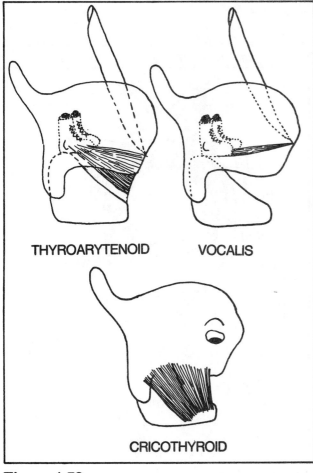

Figure 4-73

Fig. 4-74. Relationship of the vocalis and thyroarytenoid muscles to the vocal fold (coronal section). Compare with figure 4-70. The vocal muscles move the vocal fold (the true cord) whereas the **vestibular fold (the false cord) contains no muscles** and does not function significantly in pitch changes.

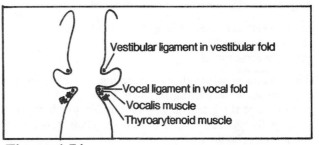

Figure 4-74

Fig. 4-75. Muscles closing the entrance to the larynx. The **thyroepiglottic muscle** lies in the aryepiglottic fold. It connects with and **pulls down the epiglottis** (like a barrel cover closing). The **aryepiglottic muscle is an** extension of the oblique arytenoid muscle. The aryepiglottic muscle, like the thyroepiglottic muscle, lies in the aryepiglottic fold. When the aryepiglottic muscle contracts, this **brings the aryepiglottic folds together, like a purse string closing a purse,** thus preventing food from entering the trachea. Of course, closing the glottis by moving the vocal cords together will also block entry, but this is inefficient and likely to result in a fit of coughing.

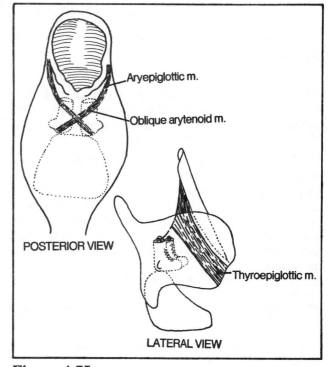

Figure 4-75

Muscles of swallowing

tongue muscles
palate muscles
hyoid muscles
pharyngeal muscles

Fig. 4-76. The intrinsic tongue muscles.

(1) superior and inferior longitudinal muscles
(2) vertical muscles
(3) transverse muscles

The tongue is talented in performing a variety of fine movements while staying in place. These movements are the result of the intrinsic tongue muscles which run longitudinally, vertically, and crosswise. These muscles do not attach to bone.

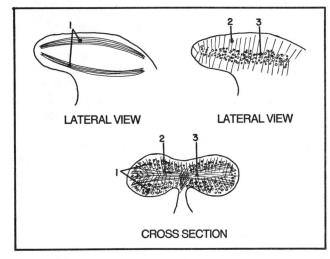

LATERAL VIEW   LATERAL VIEW

CROSS SECTION

Figure 4-76

Fig. 4-77. Extrinsic tongue muscles. Each of these muscles **attaches at one end to the tongue and at the other to a specific bone** - either the anterior mandible, the hyoid bone, or the styloid process.

(1) genioglossus (originates at the mandible and hyoid bone) The anterior fibers retract the tongue. The posterior fibers protrude the tongue ("pull it out"). If the left arm of someone riding a bicycle is paralyzed, then pushing on the right handlebar with the right hand will cause the wheel to deviate to the left. Similarly, paralysis of the left tongue results in the tongue deviating to the left on protrusion.
(2) hyoglossus (attaches to hyoid bone) - moves tongue down
(3) styloglossus (attaches to styloid process) - retracts and elevates tongue

Thus, there are muscles to move the tongue backward, forward, down, or up.

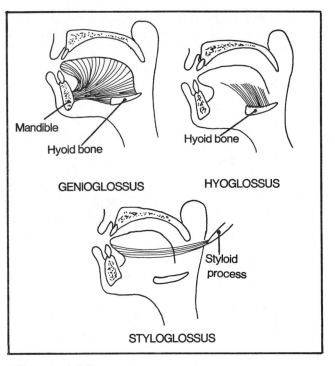

Mandible
Hyoid bone
Hyoid bone
GENIOGLOSSUS   **HYOGLOSSUS**
Styloid process
STYLOGLOSSUS

Figure 4-77

Fig. 4-78. The four paired muscles of the soft palate, as seen from the pharynx (compare with fig. 4-69). The

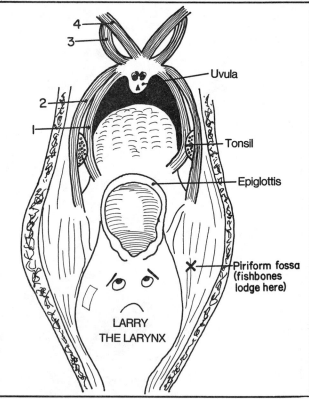

4
3
Uvula
2
1
Tonsil
Epiglottis
Piriform fossa (fishbones lodge here)
LARRY THE LARYNX

Figure 4-78

palatine muscles resemble an 8-legged spider that is about to pounce upon the larynx. (Note: the faces drawn in this figure do not exist in reality).

(1) palatoglossus - elevates the tongue and narrows the fauces (the passageway from the mouth to pharynx)
(2) palatopharyngeus - narrows the fauces, elevates the pharynx
(3) tensor veli palatini (tensor palati) - anchors near the auditory tube. It tenses the palate and causes opening of the auditory tube while swallowing. It also helps close the nasopharynx in swallowing.
(4) levator veli palatini (levator palati) - anchors near the auditory tube. It elevates the soft palate and helps close the nasopharynx during swallowing.

There is also a small muscle in the uvula (**musculus uvulae**, not shown) which helps raise the uvula and close the nasopharynx while swallowing.

Fig. 4-79. The fauces. The fauces of the mouth contain two prominent pillars - the **palatoglossal arch** (contains the palatoglossal muscle) and the **palatopharyngeal arch** (contains the palatopharyngeus muscle). Notice the **palatine tonsils** hiding between the arches. Not shown are the **lingual tonsils** which lie on the very posterior aspect of the tongue, and the **pharyngeal tonsils** (adenoids - figs. 8-1 and 8-2), which lie in the nasopharynx. These require a mirror for examination. The palatine, lingual, and pharyngeal tonsils are lymphoidal tissues which collectively are called the **tonsillar ring**.

Fig. 4-80. Muscles of the pharynx

(1) superior constrictor
(2) middle constrictor
(3) inferior constrictor
(4) stylopharyngeus
(5) salpingopharyngeus

The three constrictor muscles of the pharynx resemble 3 stacked cups. The two elevator muscles resemble straws fitting into the cups. All the constrictors attach posteriorly on a vertical midline connective tissue raphe. Anteriorly, each of the constrictors attaches to its favorite structure, which appears more likely to be bone the higher the muscle lies. For instance, the superior constrictor chooses 2 bones - mandible, and medial pterygoid plate of the sphenoid bone. The middle constrictor chooses only one bone - the hyoid bone. The inferior constrictor connects not with bone but with the thyroid and cricoid cartilages. Note that the pterygomandibular raphe forms a common origin for the superior constrictor and buccinator muscles (also see fig. 4-57).

The constrictor muscles constrict the pharynx. The stylopharyngeus elevates the pharynx. The salpingopharyngeus (which extends all the way up to the auditory tube) elevates the nasopharynx.

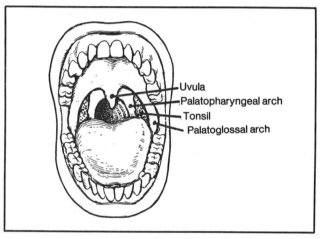

Figure 4-79

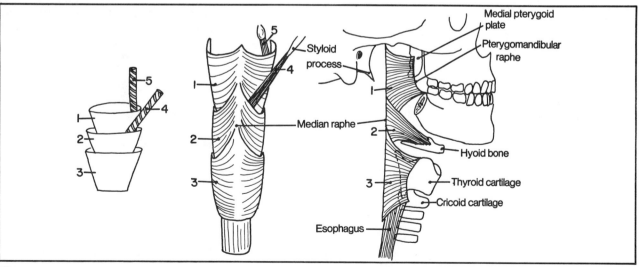

Figure 4-80

Hyoid muscles

The hyoid is a free-floating bone that provides an anchor for various muscles. **Muscles above the hyoid bone** tend to **raise the hyoid bone**, which then pushes against and raises the tongue and floor of the mouth. **Hyoid muscles below the hyoid bone** tend to **lower this bone** along with the larynx.

Fig. 4-81. The suprahyoid muscles.

(1) mylohyoid
(2) geniohyoid (dotted lines) - the only suprahyoid muscle to lie in the mouth (just deep to the mylohyoid)
(3) digastric
(4) stylohyoid

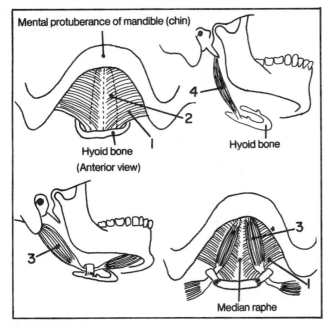

Figure 4-81

If you were a suprahyoid muscle which bone would you attach to, apart from the hyoid bone? The other end of the muscle has the option of attaching either to mandible or to temporal bone. The mylohyoid and geniohyoid attach to the mandible. The stylohyoid attaches to the styloid process of the temporal bone. The digastric ("two bellies") can't make up its mind and attaches to two bones - mandible and temporal.

Fig. 4-82. The infrahyoid muscles (see also fig. 17-4). Question: Which of the infrahyoid muscles differs basically from the others? Ans. Sternothyroid - it doesn't connect with the hyoid bone, although it is still below the hyoid. It lowers the thyroid cartilage and larynx. The omohyoid is also somewhat peculiar. It connects way out in Omaha, Nebraska, with the scapula.

The Mechanics Of Swallowing

In swallowing, the food has to reach the esophagus while avoiding both the nasopharynx (which would cause food to exit through the nose) and the larynx (which would induce a bout of violent coughing).

Fig. 4-83. Phases of swallowing. The pharyngeal wall has been opened posteriorly. A pill standing on the tongue is peering down into the pharynx.

Phase 1 - voluntary. The tongue rises against the hard palate and pushes the pill into the pharynx. The palatoglossus(1) and palatopharyngeus(2) muscles constrict the fauces, sealing the exit back to the mouth.

Phase 2 - involuntary. Food is prevented from entering the nasal cavity by contraction of the levator palati(3) and tensor palati(4) muscles, which pull the soft palate up and back against the posterior wall of the pharynx. The superior(5), middle(6), and inferior(7) pharyngeal constrictor muscles constrict sequentially, pushing food into the esophagus. At the same time the other pharyngeal muscles - palatopharyngeus(2), salpingopharyngeus(8),

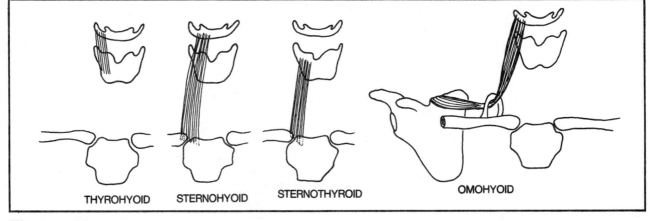

Figure 4-82

and stylopharyngeus(9) pull up the walls of the pharynx and attached larynx so that the larynx is moved against the epiglottis, thereby helping seal off the passageway to the larynx. Another reason why the larynx elevates is that the hyoid bone (not shown), which connects by muscles to the larynx, elevates in the act of pushing up the tongue.

The larynx is further sealed off by action of the thyro-epiglottic and aryepiglottic muscles (fig. 4-75).

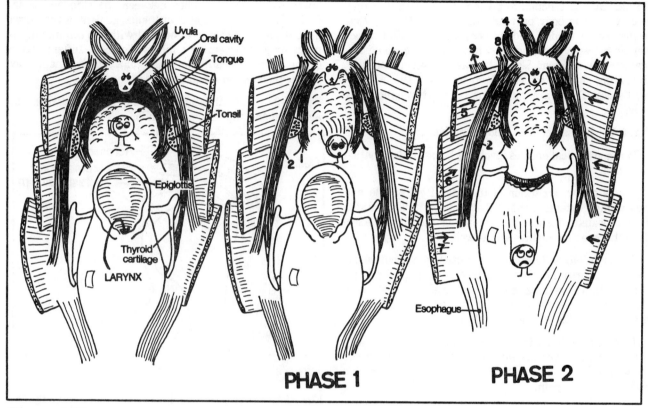

Figure 4-83

CHAPTER 5. BURSAE AND SYNOVIAL SHEATHS

Bursae

A bursa is a collapsed sac that contains a small amount of fluid. Bursae function to prevent the friction that may occur when one body part rubs on another. Usually they lie between a tendon and its bony insertion. Sometimes they lie between two tendons, which otherwise would rub against one another. Some bursae lie subcutaneously, preventing skin movements from irritating the underlying tissues. There are numerous bursae throughout the body.

Fig. 5-1. Bursa between tendon and bone.

Common clinical entities involving bursal inflammation (bursitis):

1. Tennis elbow. Irritation of the lateral epicondyle and the common extensor tendon that ends on it. This results from repeated extensor movements of the elbow. Associated bursae may be involved as well.

2. Student's elbow (olecranon bursitis). A subcutaneous olecranon bursa lies between olecranon and overlying skin. Continued resting of the elbow, e.g. on a desk, may cause friction between the skin and underlying tissues, leading to bursitis with local swelling.

3. Trochanteric bursitis. The trochanteric bursa lies deep to the gluteus maximus muscle and its insertion on the lateral aspect of the greater trochanter. Bursitis in this area causes local pain.

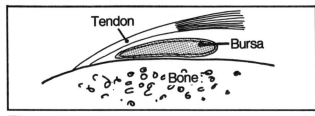

Figure 5-1

4. Calcaneal bursitis. This affects the bursa between the achilles tendon and the calcaneus. This bursitis is most common in individuals who plantar flex the ankle repeatedly, such as dancers and long distance runners.

5. Bursitis of the shoulder. A subacromial bursa lies under the acromion and rests on the shoulder joint capsule and supraspinatus tendon (fig. 5-2). This bursa may be irritated by movements of the shoulder, particularly abduction.

6. Bursitis of the knee. In prepatellar bursitis (housemaid's knee), the affected bursa lies between the skin and underlying patella. Bursitis with local swelling may occur when too much time is spent resting on the knees, as in scrubbing floors. Depending on the angle of resting, there may instead be bursitis of the subcutaneous infrapatellar bursa (fig. 5-3) which lies between the skin and tibia. **Baker's cyst** is a bursitis with enlargement of the bursa behind the knee that is associated with the semimembranosus or gastrocnemius tendons (fig. 5-3). Inflammation of the bursa beneath the tibial collateral ligament may induce pain on the medial aspect of the knee.

Fig. 5-2. The subacromial bursa. It lies between the acromion and supraspinatus tendon.

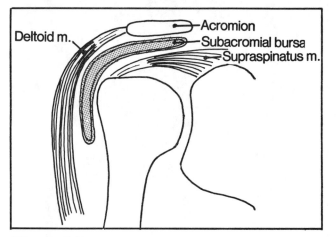

Figure 5-2

Fig. 5-3. Bursae of the knee that are commonly involved in bursitis.

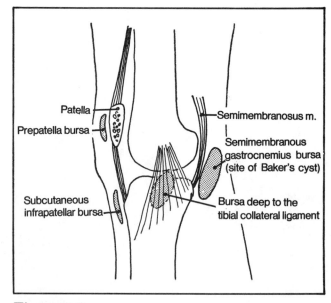

Figure 5-3

Synovial Sheaths

Synovial sheaths are similar to bursae. They differ in that rather than being rounded flat sacs, they are tubular structures which surround long tendons, following them over long distances and on all sides. Synovial sheaths are mainly found in the hand and foot, which contain long tendons.

Fig. 5-4. A typical synovial sheath. Blood vessels enter by way of the mesotendon. Within the fingers and toes, the mesotendon is evident only in spotty areas (**vinculae**) where blood vessels enter. In a sense, the mesotendon is analogous to mesentery, through which certain visceral organs (e.g., the small intestine) receive their blood supply (fig. 9-13). A fibrous tunnel may surround and anchor the synovial sheath and tendon to the bone and prevent bowstringing during tendon movements.

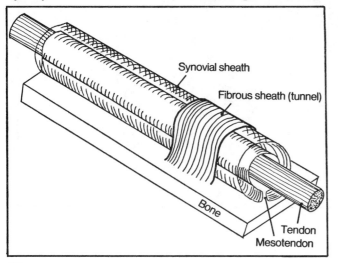

Figure 5-4

Fig. 5-5. The disposition of synovial sheaths in the palmar aspect of the hand. A, synovial sheaths (shaded); B, fibrous sheaths. These (B) anchor the synovial sheaths and tendons, and prevent bowstringing. The palmar aponeurosis has a similar function in the palm, and the flexor retinaculum has a similar function more proximally, near the wrist.

An infection within a synovial sheath may extend along its length. Thus, infections within the synovial sheath of the thumb or small finger are more likely to reach the wrist than infections of the middle 3 fingers.

Corresponding to the flexor retinaculum, an extensor retinaculum is found on the dorsum of the hand. It lies a little more proximal than the flexor retinaculum, part of it connecting with the radius and ulna. It encloses the long extensor tendons. The synovial sheaths of the long extensor tendons (not shown in fig. 5-5) lie largely under the extensor retinaculum and do not extend along the phalanges as do the flexor synovial sheaths (see fig. 5-5C).

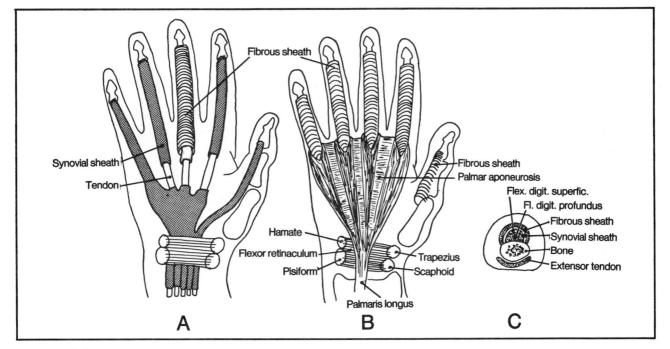

Figure 5-5

Fig. 5-6. Synovial sheaths of the dorsal foot. The locations of dorsal and plantar synovial sheaths of the foot roughly correspond to those in the hand; i.e., the dorsal sheaths do not extend along the phalanges, whereas the plantar sheaths (not shown) do. Do not confuse retinacula with bony ligaments, which lie deep to the tendons.

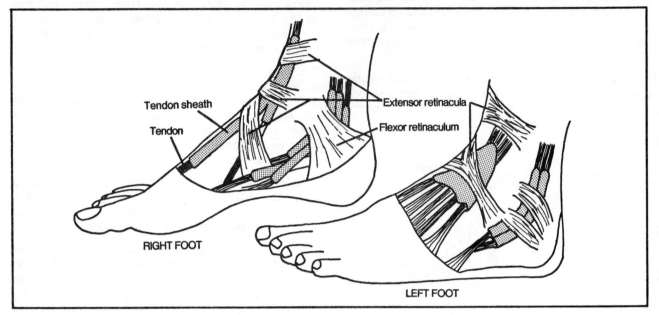

Figure 5-6

CHAPTER 6. THE CIRCULATORY SYSTEM

The Heart

Fig. 6-1. Summary of the circulation through the major chambers of the heart.

The term "base of the heart" frequently occurs in the medical literature. It refers to the posterior surface of the heart, **not** the diaphragmatic surface; the base is not illustrated in figure 6-1.

In right heart failure, fluid backs up and collects in the liver and legs. In left heart failure, fluid collects in the lungs (**pulmonary edema**).

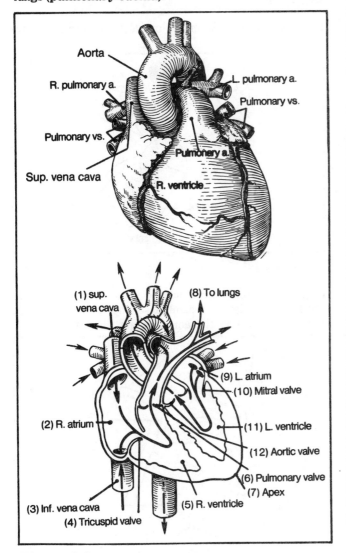

Figure 6-1

Fig. 6-2. Schematic view of the right atrium (plus a few other things).

(1) smooth walled area of right atrium
(2) rough walled area of right atrium
(3) opening of inferior vena cava into smooth walled area
(4) opening of superior vena cava into smooth walled area
(5) opening of coronary sinus into smooth walled area
(6) fossa ovalis
(7) sinoatrial (SA) node
(8) atrioventricular (AV) node
(9) tricuspid valve
(10) chorda tendinae
(11) papillary muscle

Once upon a time a cherub who lived in the liver entered the hepatic veins, which brought him to the heart via the inferior vena cava. The vena cava enters the heart through the floor of the right atrium. On entry into the right atrium the cherub noted that the atrium consisted of two main parts (fig. 6-2):

1. A smooth-walled posterior part where he entered.
2. A rough-walled trabecular part, anteriorly.

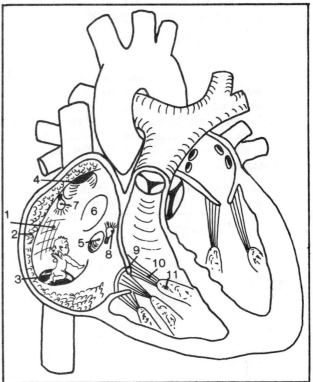

Figure 6-2

The smooth-walled part had other points of entry besides that of the inferior vena cava: a hole in the roof (superior vena caval entrance) and a hole in the medial wall (entry point of the coronary venous sinus, which drains the heart itself). With all these jet streams hitting him in the smooth walled area, the cherub tried to escape that area. He blindly swam about, bumping his head, unfortunately, on the medial wall of the smooth-walled chamber, forming an indentation (fossa ovalis) posterior to the coronary sinus opening. It was too bad that he didn't strike his head harder; he would then have broken through the wall and escaped into the left atrium (an abnormal congenital opening of this sort is called a **patent foramen ovale**). It was quite dark in the smooth-walled right atrium. The cherub reached up to turn on a ceiling light (SA node) near the opening of the superior vena cava and a wall light (AV node) near the opening of the coronary sinus. These light bulb nodes actually lie deeply, embedded in the atrial wall and are part of the electrical conduction system of the heart (fig. 6-6).

With the light on, the cherub recognized that the best exit (the tricuspid valve) lay in the rough-walled (trabecular) zone of the atrium. He stepped across the threshold between smooth and trabecular areas and fell through the tricuspid valve into the right ventricle. The tricuspid valve in a vague sense resembles a parachute that contains 3 chutes (cusps), strings (chorda tendinae) and a load at the ends of the strings (papillary muscles). The mitral (bicuspid) valve in the left ventricle appears somewhat similar but contains only two cusps.

Fig. 6-3. The tricuspid valve (laid out flat) resembles a parachute.

On ventricular contraction, the cusps of the tricuspid valve moved together and prevented the cherub from backflowing from the right ventricle into the right atrium. The **chorda tendinae** prevented the cusps from flopping back into the atrium. The pulmonary and aortic semilunar valves each have three cusps which do not have chorda tendinae.

The little cherub was propelled through the lungs and heart chambers into the aorta. He noticed, from the

vantage point of the aortic arch, that the semilunar valves of the aorta and pulmonary artery are arranged facing one another in a romantic mirror image configuration (fig. 6-4).

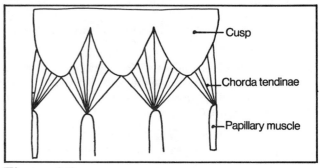

Figure 6-3

Fig. 6-4. The aortic and pulmonary valves as seen from above.

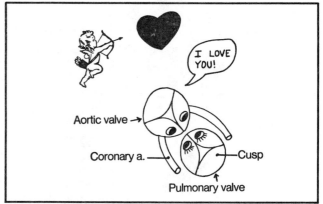

Figure 6-4

Fig. 6-5. Coronary arterial and venous circulation. The coronary circulation resembles a modified jock strap. Blood, on reaching the aorta, enters the right and left

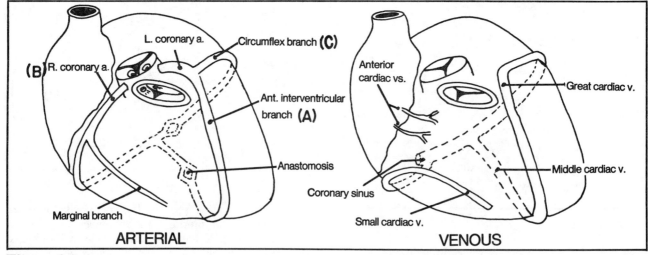

Figure 6-5

coronary arteries. The two coronary arteries encircle the heart like a waist band. The left coronary artery gives off an **anterior interventricular** (left anterior descending) **branch**, which also travels around the heart. The right coronary artery gives off a **marginal branch**. There are significant variations to this scheme, especially regarding exact sites of anastomoses. For instance, the right coronary artery commonly is "dominant" when it crosses to the left side to anastomose posteriorly with the left coronary artery. Sometimes the left coronary artery is dominant. Dominance is important as it determines which area of the cardiac wall is damaged during a myocardial infarction (heart attack). A, B, and C in figure 6-5 indicate the most commonly occluded vessels in a myocardial infarction, in decreasing order of occurrence. Commonly, occlusion occurs near the origin of the respective vessels.

The **coronary venous circulation** looks somewhat similar to the arterial circulation. Imagine that the first segments of the right and left coronary arteries are removed, as the veins do not connect with the aorta. In their place, introduce, posteriorly, a connection with the right atrium, where **the coronary sinus** drains into the right atrium. Also add a couple of **anterior cardiac veins**. **Thesbian veins** (not shown) directly drain from the cardiac muscle (myocardium) into the cardiac chambers, especially into the atria.

Fig. 6-6. The conduction system of the heart. This consists of modified muscle cells that relay electrophysiological impulses through the heart. The SA node lies within the right atrium and is the main pacemaker,

firing on the average of about 72 times per minute. Impulses spread throughout the right and left atrial musculature , inducing atrial contraction. The impulses arrive at the AV node, which relays the impulses to the **bundle of His** (AV bundle) which runs in the interventricular septum as two bundles, one to the right ventricle and one to the left ventricle.

Interference with the blood circulation to the AV node or AV bundle may prevent SA nodal impulses from reaching the ventricular system (**heart block**). Blockage at a more distal level of the conduction system may result in **right or left bundle branch block**. Such conditions are detectable on an electrocardiogram.

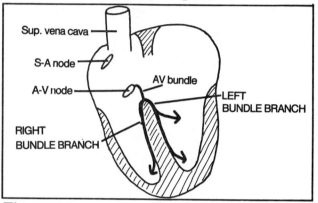

Figure 6-6

Fig. 6-7. The cardiac silhoutte. Note the boundaries of the heart. These are important in examining x-rays of the chest, in which only the silhouette is clearly seen.

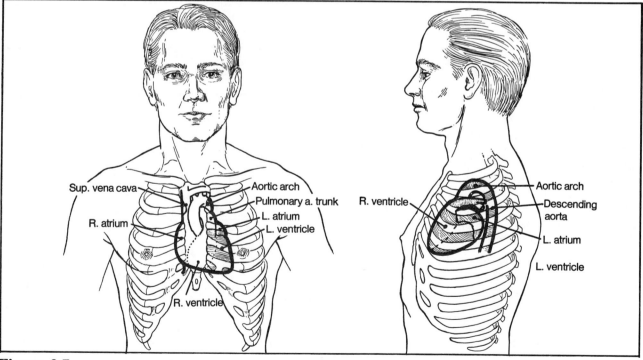

Figure 6-7

Fig. 6-7A. Positions of the heart sounds. Cardiac valve murmurs are heard best not directly over each valve but somewhat in advance of the valve, where blood passes beyond the valve and comes closest to the chest wall:

M, position for listening to mitral valve sounds
T, position for tricuspid valve sounds
A, position for aortic valve sounds
P, position for pulmonary valve sounds

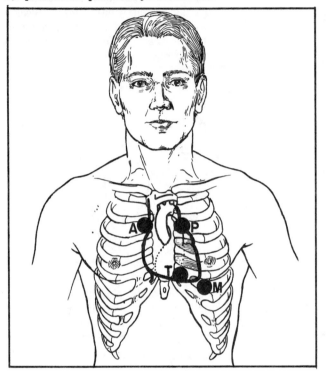

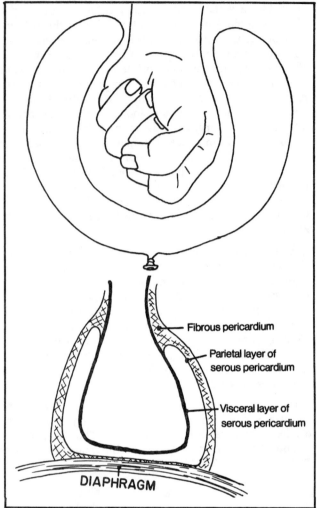

Figure 6-8

Fig. 6-8. The pericardium. The **serous pericardium** is a membranous covering of the heart, arranged somewhat like a balloon surrounding an indenting fist. Normally the cavity of the serous pericardium (the inside of the balloon) contains a thin layer of pericardial fluid, enabling the heart to beat and move in a relatively frictionless manner. Surrounding the serous pericardium is the **fibrous pericardium.** The fibrous pericardium anchors to the diaphragm and to the sternum and extends up a short distance along the great vessels. The **internal thoracic artery** (fig. 6-29) supplies much of the **fibrous** and **parietal pericardium.** The **coronary arteries supply the visceral pericardium.**

The parietal and fibrous pericardia are tightly apposed. Fluid may collect between the parietal and visceral pericardia during inflammation (**pericarditis**) and may have to be removed if it threatens to compress the heart (**cardiac tamponade**) and inhibit cardiac function.

Fig. 6-9. Positioning of the syringe in drainage of a pericardial effusion. This method minimizes the risk of puncturing a lung or a major coronary vessel.

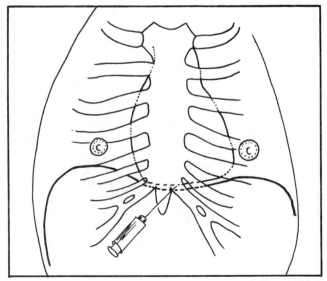

Figure 6-9

Fig. 6-10. The oblique and transverse sinuses. The anterior part of the pericardial sac has been removed along with the entire heart. Note that all veins (superior and inferior vena cava; pulmonary veins) are collectively outlined by one portion of serous pericardium. The arteries (aorta and pulmonary artery) are outlined by another portion of serous pericardium. Thus, a cardiac surgeon wishing to gain access to the base of the aorta or pulmonary artery may do so by opening the pericardium and passing a finger around the aorta and pulmonary artery via the transverse sinus. Approaching the transverse sinus from below, however, via the oblique sinus, would be unsuccessful as the way is obstructed by folds of pericardium.

Fig. 6-11. Cardiac circulation (schematic view) in the fetus and postnatal. As in the adult, all arteries in the fetus convey blood away from the (fetal) heart, whereas veins convey blood toward the (fetal) heart. Since nutrition is supplied by the mother, there is no need for blood to enter its lung for oxygenation. Thus, blood is largely shunted past the lungs. Blood avoids entry to the lung by passing from the right atrium to the left atrium through the **foramen ovale,** a hole in the interatrial wall. Blood may pass directly from pulmonary artery to aorta by the **ductus arteriosus.** Normally the foramen ovale and ductus arteriosus close after birth.

The interatrial wall of the fetus contains a **septum primum,** which acts as a valve that presses against the **septum secundum.** The valve-like arrangement allows blood to pass from right to left rather than left to right. Normally, the septum primum and septum secundum fuse after birth, leaving an indentation called the **fossa ovalis.**

Numerous variations may occur in the development of the atrial and ventricular partitions, in the valves, and in the positions of the great vessels. In atrial septal defects, an opening persists between the right and left atria. Commonly this involves a deficit in the septum primum or septum secundum. Small defects may be inconsequential.

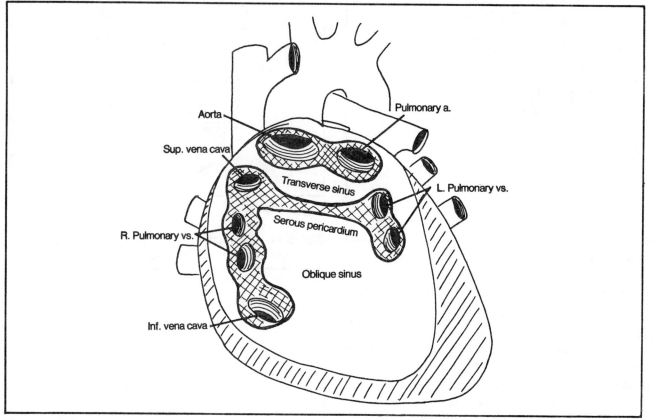

Figure 6-10

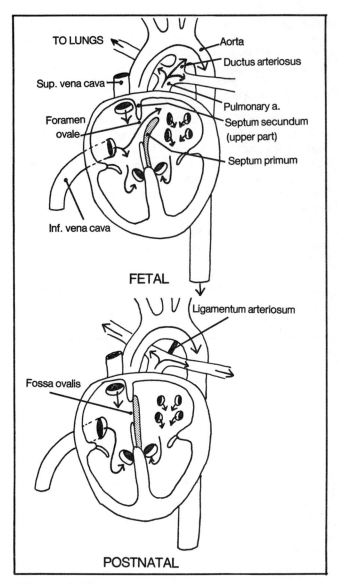

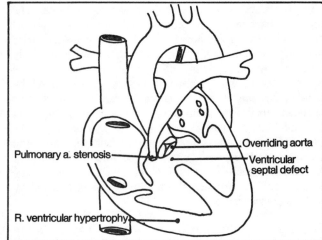

Figure 6-12

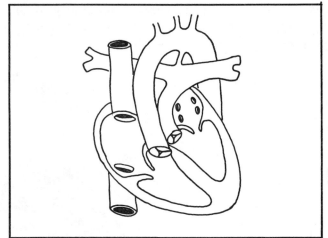

Figure 6-13

Figure 6-11

Fig. 6-14. Coarctation (focal narrowing) of the aorta. Circuitous vascular channels (see fig. 6-29) may be called

Fig. 6-12. The tetralogy of Fallot. Classically, this congenital cardiac anomaly includes 4 components: 1) Interventricular septal defect. 2) Pulmonary valve narrowing (stenosis). 3) Overriding aorta (the opening of the aorta is shifted to override the septal defects. 4) Thickening (hypertrophy) of the right ventricle.

Fig. 6-13. Transposition of the great arteries. The aorta originates from the right ventricle and the pulmonary artery originates from the left ventricle. This defect is incompatible with life unless there is an anastomotic shunt, such as a **patent ductus arteriosus**, in which the connection between aorta and pulmonary artery persists.

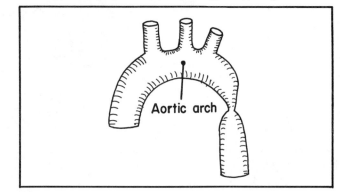

Figure 6-14

upon to divert blood from the aortic arch to the lower extremities. When this involves dilation of the intercostal arteries, there may be notching of the ribs on x-ray examination.

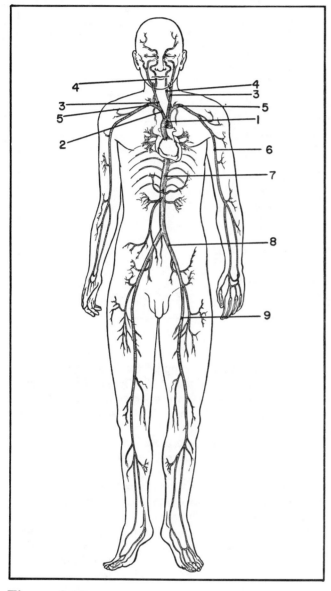

Figure 6-15

Fig. 6-15. Overview of the arterial system.

(1) aortic arch
(2) brachiocephalic (innominate) a.
(3) common carotid a.
(4) vertebral a.
(5) subclavian a.
(6) brachial a.
(7) descending aorta
(8) common iliac a.
(9) femoral a.

Fig. 6-16. The internal carotid artery. Notice how simple it is to remember the branches coming off certain arteries. For instance:

1. The brachiocephalic artery has no branches from beginning to end.
2. The common carotid artery has no branches from beginning to end.
3. The internal carotid artery (called internal because if extends internally, into the skull, to supply the brain) has one key branch before ending - the **ophthalmic artery**, which supplies the retina and surrounding tissues of the orbit. There are also several tiny branches (not elaborated upon here) which become important to the neurosurgeon as alternate anastomotic blood routes when they dilate in response to tumors or vascular compromise.

At the junction between internal and external carotid arteries (about the level of the superior border of the thyroid cartilage) there lie a **carotid body** and a **carotid sinus**. The carotid body is a chemoreceptor that is sensitive to oxygen and carbon dioxide levels in the blood and assists in the control of respiration. The carotid sinus is a pressure sensitive zone on the bifurcation that assists in regulating blood pressure and pulse rate. Massage of the carotid sinus leads to slowing of the heart rate, a measure that can be useful in the treatment of certain rapid cardiac rhythms (**atrial tachycardia**).

The external carotid artery runs external to the skull and has many branches to compensate for the paucity of branches from the internal carotid artery.

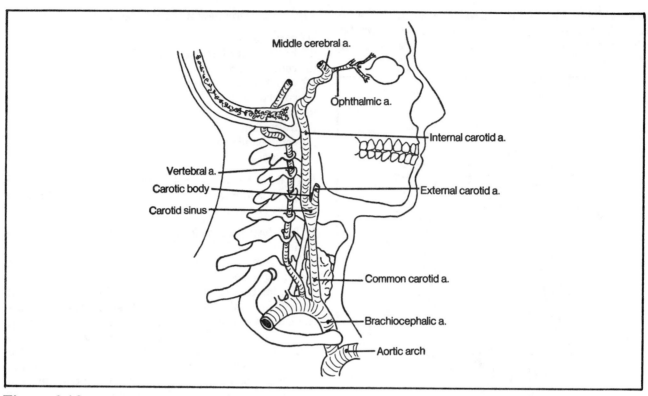

Figure 6-16

Fig. 6-17. Branches of the external carotid artery. As the bifurcation of the common carotid artery occurs at about the level of the angle of the jaw, most branches of the external carotid artery lie above this level and hence supply the face rather than the neck. An exception is the first branch, the superior thyroid artery which falls down on the job. It is afraid of heights and reaches down to grab the thyroid gland.

(1) superior thyroid a. - supplies the thyroid gland and larynx

(2) ascending pharyngeal a. To compensate for the superior thyroid artery's going down in such an embarrassing manner, the next branch, the ascending pharyngeal artery, reaches high to the base of the skull, giving off a branch to the pharynx as well as the posterior meningeal artery to supply the meninges.

(3) lingual a. - dives deep under the mandible to supply the tongue. It is helpful to know that the lingual artery originates posterior to the end of the hyoid bone. This is an important surgical landmark as it sometimes is necessary to tie off the lingual artery in surgical procedures involving the tongue and nearby structures.

(4) facial a. This artery proceeds in a somewhat drunken manner. It starts to go deep to the mandible and then says "Hey, what am I doing here? The lingual artery has already passed this way. I should be supplying new territory." So the facial artery moves back around the mandible and externally supplies the face, staggering in an irregular manner until it finally reaches the eye, where it anastomoses with the ophthalmic artery. Maybe that's partly why the eyes get bloodshot when one is drunk.

(5) occipital a. - goes to the occipital area

(6) posterior auricular artery - goes to the posterior auricular area

(7) superficial temporal a. It may be biopsied as part of the diagnosis of **temporal arteritis**, an inflammatory condition affecting elderly people.

(8) maxillary a. - goes to a maximum number of places. It arises near the temporomandibular joint, runs into the pterygopalatine fossa (behind the maxillary bone) and, like the ophthalmic and facial arteries, it has a predisposition to end up in the orbit, as the infraorbital artery (8d). Along the way, though, it gives off some rather bloody branches, which often are responsible for various clinical conditions:

(8a) middle meningeal a. - supplies the meninges and may bleed in an epidural hemorrhage

(8b) sphenopalatine a. - responsible for some severe nosebleeds

(8c) inferior and superior alveolar as. - supply the teeth.

The maxillary artery does not have a monopoly on hemorrhage production. The facial artery gives off a tonsillar branch that may bleed severely in a tonsillec-

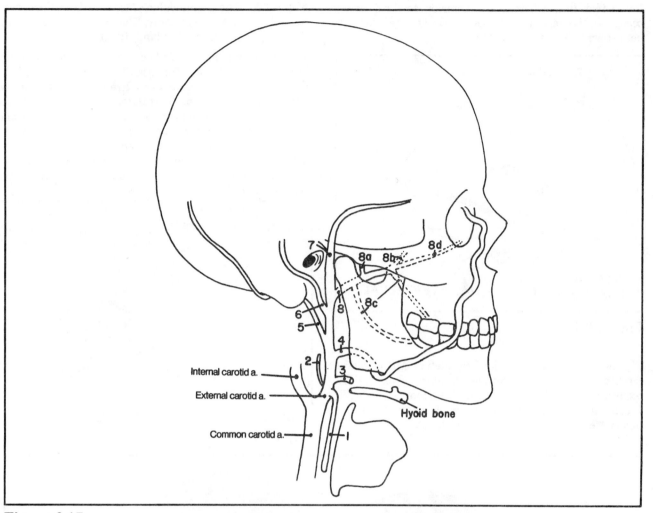

Figure 6-17

tomy (although it is usually venous bleeding that causes difficulty in this procedure).

Note that the exact sequence of these arteries may vary. They can be remembered by figuring out the general regions of the head that need to be vascularized.

Arteries Of the Brain (internal carotid and vertebral arteries)

Fig. 6-18. The major arterial supply to the brain. ACA, anterior cerebral artery; MCA, middle cerebral artery; PCA, posterior cerebral artery; PAD, pia, arachnoid, dural membranes (they PAD the brain).

There are two main arteries that supply the brain - the vertebral and internal carotid arteries. Review the general topography of the central nervous system in figure 12-1. The vertebral artery changes its name. It's called the basilar artery at the level of the pons and the posterior cerebral artery at the level of the cerebrum (fig.

6-18). You'll see why when we discuss Willis, a ferocious spider that lives in the brain.

Note the important imaginary line in figure 6-18. It divides the cerebrum into a front (anterior) and a back (posterior) area. The internal carotid artery supplies the front area. Obstruction of the right carotid artery causes weakness and loss of sensation on the left side of the body (one side of the brain connects with the opposite side of the body and the area of the brain above the dotted line is involved with such movement and sensation). Blockage of the circulation under the dotted line affects the circulation to the visual area of the cerebrum, the brain stem, and the cerebellum and may result in visual loss, dizziness, and other problems.

The internal carotid artery divides into an anterior and a middle cerebral artery. Note (fig. 6-19) that the posterior cerebral artery occupies the entire cerebrum below the dotted line. The middle cerebral artery, though, occupies only the lateral surface of the cerebrum above

79

the dotted line, whereas the anterior cerebral artery occupies the entire midline area of the cerebral hemisphere above the dotted line.

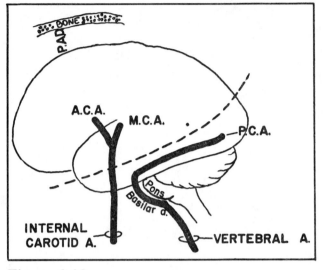

Figure 6-18

Fig. 6-19. The cerebral blood circulation. MCA, middle cerebral artery; ACA, anterior cerebral artery; PCA, posterior cerebral artery. (Modified from DeMyer,W., Technique of the Neurologic Examination, with permission of McGraw-Hill Book Company, 1974).

Fig. 6-20. The humunculus. (Modified from Carpenter, M.B., Human Neuroanatomy, The Williams and Wilkins Company, Baltimore, Maryland, 1982).

The brain contains an upside down man named HAL (H-head, A-arm, L-leg), functionally represented on the cerebral cortex. HAL's lower extremity bends over the top of the cerebrum. Therefore, an occlusion of the anterior cerebral artery results in loss of strength and sensation in the lower part of the body, as the area of the brain it supplies is more concerned with the lower extremities. An occlusion of the middle cerebral artery predominantly affects strength and sensation in the upper regions of the body.

Figure 6-20

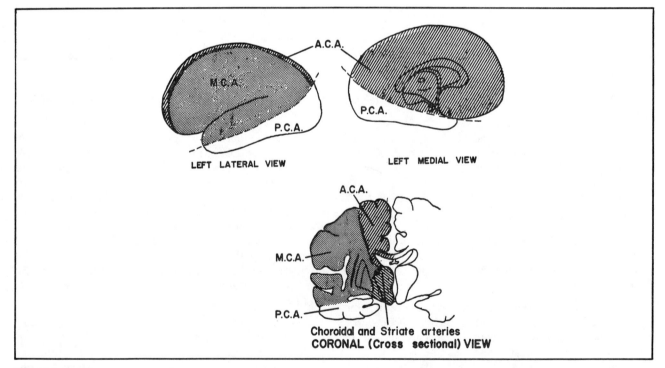

Figure 6-19

Fig. 6-21. Willis! (from Goldberg, S., Clinical Neuroanatomy Made Ridiculously Simple, Medmaster, 1983).

A ferocious spider lives in the brain. His name is Willis! Note (fig. 6-21) that he has a nose, angry eyebrows, two suckers, eyes that look outward, a crew cut, antennae, a fuzzy beard, 8 legs, a belly that, according to your point of view, is either thin (basilar artery) or fat (the pons, which extends from one end of the basilar artery to the other), two feelers on his rear legs, and male genitalia. The names in figure 6-21 look similar to those in figure 6-18 because they are the same structures, seen from different angles. In figure 6-21 the brain is seen from below, so the carotid arteries are seen in cross section. Figure 6-21 also explains why the vertebral artery changes its name twice. At first the two vertebral arteries fuse to form one basilar artery. The basilar artery then divides again into two posterior cerebral arteries.

An occlusion of the basilar artery at the junction of the two posterior cerebral arteries will result in total blindness, as the posterior cerebral arteries supply the visual area of the cerebrum (fig. 6-18). Occlusion of a vertebral artery may result in little or no deficit because of the remaining blood supply from the opposite vertebral artery.

The two communicating arteries are shown as dotted line in figure 6-21 because blood flow shows no particular tendency to go one way or the other along these channels. This is logical since blood normally flows up both the carotid and vertebral arteries, equalizing the pressure between the two circulations. Hence, contrast material (for x-ray studies) injected into the right carotid artery generally will not flow back into the basilar artery across the posterior communicating artery. Nor will it usually cross over to the left side of the brain via the anterior communicating artery. This all goes to show that the brain is smart. If one of the major vessels is occluded, the communicating arteries function as anastomoses.

Willis (fig. 6-21) has hairy armpits - the third cranial nerve exits between the posterior cerebral artery and the superior cerebellar artery. An **aneurysm** (a weakness and focal ballooning out of the wall of a blood vessel) which affects either of the above two blood vessels may press upon and damage the third nerve. Willis' head (the **Circle of Willis**) is a common area for aneurysms.

There is an anterior, middle, and posterior cerebral artery. It would have been nice to have an anterior, middle, and posterior cerebellar artery, too, but some inconsiderate SAP named these arteries after himself (S-superior cerebellar artery, A-anterior inferior cerebellar artery, P-posterior inferior cerebellar artery). The cerebellar arteries supply not only the cerebellum but also parts of the brain stem. Their occlusion will result in damage to corresponding areas of the brain stem.

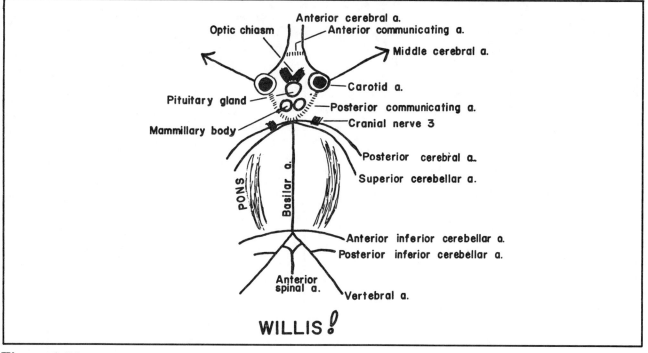

Figure 6-21

Fig. 6-22. The internal carotid and vertebral arteries. Olf., olfactory tract; MS, medial striate a.; LS, lateral striate a.; ACh., anterior choroidal a.; MC, middle cerebral a.; 3, cranial nerve 3; mb, midbrain; IC, internal carotid a.; V, vertebral a.; MED, medulla; EC, external carotid a.; CC, common carotid a.; S, subclavian a.; A, aorta. The internal carotid artery enters the interior of the skull by the **foramen lacerum** (fig. 2-27). The

radiologically by injecting an artery with a contrast material that will outline the blood vessels on x-ray film. This will reveal whether the vessel is blocked, leaking, or of abnormal form or position, resulting from displacement by a tumor or hemorrhage. A catheter (injection tube) threaded retrograde up the right brachial artery to the subclavian artery at the level of the right vertebral and carotid arteries can be used to release contrast material that will enter both the right vertebral and right carotid arteries, thereby demonstrating the front and back cerebral circulations (fig. 6-18). Injection, however, on the left side would demonstrate only the posterior circulation, since the left carotid artery arises directly from the aorta (fig. 6-22). Thus, the choice of artery and side is important in showing up the desired area in x-ray.

Fig. 6-23. Arterial vascularization of the spinal cord. The anterior 2/3 of the cervical cord are supplied by the **anterior spinal artery** (fig. 6-21), whereas the posterior 1/3 is supplied by **two posterior spinal arteries** that arise variably from Willis' lower body area. The anterior and posterior spinal arteries extend the full length of the spinal cord. There are additional anastomoses from **radicular branches** that come off the **vertebral, inferior thyroid** (fig. 6-26), **intercostal** (fig. 6-30) and **lumbar** (fig. 6-31) arteries. Hence, the spinal cord has a

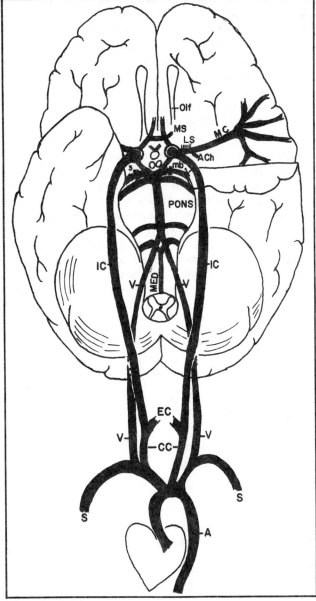

Figure 6-22

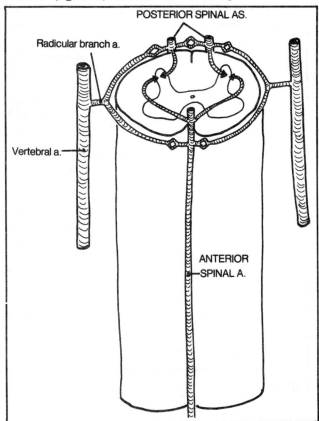

Figure 6-23

vertebral artery ascends through the transverse foramena of vertebrae C1-C6 to enter the skull via the **foramen magnum** (fig. 6-16). This route allows it to give off branches to the spinal cord and medulla.

One can study the anatomy of the cerebral circulation

rich anastomotic vascular system and only infrequently is involved in strokes. However, segments T1-T4 are relatively weak areas, more susceptible than others to infarction.

It is possible to suffer a major cerebral stroke either by a massive arterial occlusion or by an occlusion of a very small artery situated in a critical area. For instance, a sudden occlusion of the internal carotid artery may cause a massive cerebral infarction. An occlusion of one of the tiny arteries that arise from Willis' head (**medial striate, lateral striate, or anterior choroidal arteries** - figs. 6-19 and 6-22) may cause as much deficit by infarcting critical pathways between the cerebrum and the brain stem that converge just behind Willis' head (fig. 6-24).

Fig. 6-24. The internal capsule. Each cerebral hemisphere contains one internal capsule, situated deep within the brain just behind Willis' head. The internal capsule contains critical motor and sensory pathways that connect cerebrum and brain stem. The anterior choroidal artery and striate arteries supply the internal capsule and may hemorrhage in situations of hypertension or arteriosclerosis, thereby causing much damage from a tiny lesion.

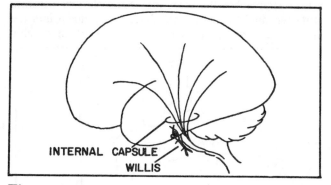

Figure 6-24

Arteries Of the Neck

The carotid and vertebral arteries mainly vascularize the head (above the level of the mandible) except for the superior thyroid artery which reaches down into the neck. The remainder of the blood supply to the neck comes mainly from other branches of the subclavian artery (figs. 6-25 and 6-26).

Fig. 6-25. The subclavian, axillary, and brachial arteries and their axillary boundaries. The **subclavian artery** extends from the brachiocephalic artery across the first rib, where it becomes the **axillary artery** (the artery of the axilla, or armpit). The subclavian artery supplies much of the neck. Hence, its branches tend to have the term "cervical" as part of their names.

The axillary artery lies between the first rib and the

humerus and ends where it passes the teres major insertion. Hence , it is in a good position to supply shoulder muscles superiorly and deeper muscles of the armpit (axilla) and lateral chest wall (the latter arteries have "thoracic" as part of the name).

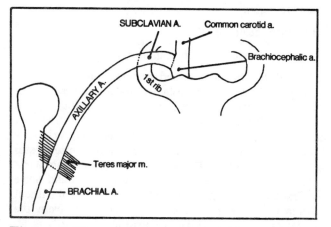

Figure 6-25

Fig. 6-26. Branches of the subclavian and axillary arteries. Branches of the **subclavian artery** may be described as Roman numeral "XI", with "XI" on top and "XI" on the bottom, the bottom "X" and "I" being somewhat farther apart than the "X" and "I" on top. Numbering in figure 6-26 is as follows:

(1) internal thoracic a. (internal mammary a.). This is the first branch of the subclavian artery and it goes totally beserk! It indirectly ends up in the groin. First it runs behind (deep to) the ribs, near the sternum, giving branches to the sternum, thoracic wall, thymus, bronchi, breast, pericardium, and diaphragm. Then it lies on the abdominal wall as the **superior epigastric artery**, all with the eventual aim of reaching the umbilicus, where anastomoses occur with the **inferior epigastric artery** (fig. 6-29). The inferior epigastric artery arises from the **external iliac artery**, runs near the inguinal ligament, is an important landmark in the surgical approach to inguinal hernias (fig. 11-5), and extends superiorly to join the superior epigastric artery. Thus, the neck connects with an inguinal hernia, the belly button being an intermediary anastomosis. This indirect routing beomes particularly important in coarctation of the aorta (fig. 6-14) where blood cannot reach the lower extremities by the normal channels.

(2) vertebral a. - already described

(3) thyrocervical trunk. Note the terms "thyro" and "cervical" implying blood supply to the thyroid gland and the cervical, or neck, region. This octopus-like artery (there are variations) has three main branches (the first letters of which spell "SIT"):

3a. Suprascapular a. - supplies muscles in the scapular region

3b. Inferior thyroid a. - reaches up to the thyroid gland, supplying muscles along the way. One of its branches, the **ascending cervical artery**, contributes branches to the spinal cord.

3c. Transverse cervical a. - supplies muscles in the scapular region

(4) costocervical trunk:

4a. deep cervical branch - supplies the back of the neck

4b and 4c. first and second posterior intercostal as. - supply the first two intercostal spaces, anastomosing with the internal thoracic artery, as do the intercostal arteries that come off the descending aorta (see fig. 6-29). The first and second posterior intercostal arteries may originate from a common trunk called the **highest intercostal artery** (the name makes sense). The descending aorta begins about the level of vertebra T4. This would leave the first two intercostal spaces (T1-T2 and T2-T3) without vascularization were it not for the branches from the subclavian and internal thoracic arteries.

Branches of the **axillary artery** are as follows (in fig. 6-26):

(5) highest (or supreme) thoracic a. - supplies local muscles

(6) thoracoacromial a. This artery made a big mistake. Like everyone else, it confused the acromion with the coracoid process. It's supposed to go to the acromion, but it originates behind the pectoralis minor muscle near this muscle's origin on the coracoid process. Realizing its mistake, this artery comes out from behind the muscle and goes to the acromion, supplying local muscles along the way.

(7) lateral thoracic artery. After passing behind the pectoralis minor muscle, the axillary artery immediately gives off the lateral thoracic artery, which supplies part of the lateral surface of the thoracic wall. But rather than take the rough road, bumping over each rib, it takes a smooth, coasting ride along the lateral edge of the soft pectoralis minor muscle, sending branches to an even softer region - the lateral aspect of the breast. Note that the blood supply to the breast comes laterally, from the lateral thoracic artery, medially from the internal mammary artery, and deep, from the third, fourth, and fifth intercostal arteries.

(8) posterior circumflex a.

(9) anterior circumflex a.

(10) subscapular a.

Once past the pectoralis minor muscle, the axillary artery approaches the teres major muscle and becomes quite terrified. It sees itself about to fall down into the arm. In an attempt to hold on for dear life it quickly

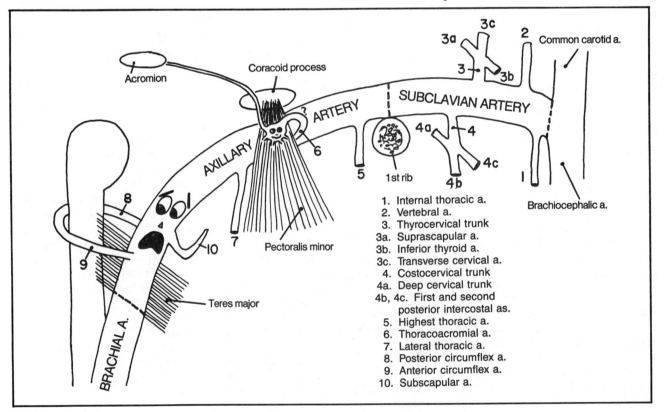

1. Internal thoracic a.
2. Vertebral a.
3. Thyrocervical trunk
3a. Suprascapular a.
3b. Inferior thyroid a.
3c. Transverse cervical a.
4. Costocervical trunk
4a. Deep cervical trunk
4b, 4c. First and second posterior intercostal as.
5. Highest thoracic a.
6. Thoracoacromial a.
7. Lateral thoracic a.
8. Posterior circumflex a.
9. Anterior circumflex a.
10. Subscapular a.

Figure 6-26

wraps one arm (anterior and posterior cimcumflex arteries) around the humerus and reaches backward with the other arm to grab the scapula (subscapular artery).

The subscapular artery anastomoses with two branches of the subclavian artery that were mentioned above - the transverse cervical and suprascapular arteries - to form an anastomotic network around the scapula. Thus, if the axillary artery were blocked, e.g. at the level of the first rib, blood could bypass the block via these anastomoses. Another anastomosis between subclavian and axillary arteries occurs between the highest thoracic and highest intercostal arteries.

If the subclavian artery is blocked just before it gives off the vertebral artery, the **subclavian steal syndrome** may develop. Blood then travels back down the vertebral artery from the brain, "stealing" blood from the basilar artery and brain stem in order to supply the upper extremity. The patient may experience dizziness or fainting with vigorous use of the affected arm.

Arteries Of the Upper Extremity

Fig. 6-27. The brachial artery. The brachial artery begins in the arm at the teres major insertion and extends below the elbow where it bifurcates as the radial and ulnar arteries. Although it is no longer the axillary artery it is still terrified, wrapping one leg (profunda brachii artery) around the humerus. Collateral branches from both the brachial and profunda brachii artery surround the elbow joint. In order to avoid loneliness, the brachial artery and its branches are accompanied by the three long nerves of the upper extremity. The **median nerve accompanies the brachial artery**; The **radial nerve accompanies the profunda brachii artery** in a spiral groove in the humerus; the **ulnar nerve accompanies the superior ulnar collateral artery**. Hemorrhage below the elbow may be controlled by direct pressure on the brachial artery, at the medial aspect of the mid-arm (fig. 6-27).

The radial artery pulse, which is felt radially in the wrist, is easier to feel than the ulnar pulse. Perhaps this is so (not really, but it helps to remember it that way) because the ulnar artery flow is made weaker by giving off the common interosseus artery.

Just as there are rich anastomoses around the elbow, which allow blood flow to proceed in the event of a major arterial obstruction, there are rich anastomoses in the wrist and hand, involving branches of the radial, ulnar, and anterior and posterior interosseus arteries. Specifically, there are four major arterial arches which interconnect with one another. Think of each of the above named arteries (ulnar, radial, anterior and posterior interosseus) as having its own arch (fig. 6-28):

Ulnar artery - **superficial palmar arch**
Radial artery - **deep palmar arch**
Anterior interosseus artery - **palmar carpal arch**
Posterior interosseus artery - **dorsal carpal arch**

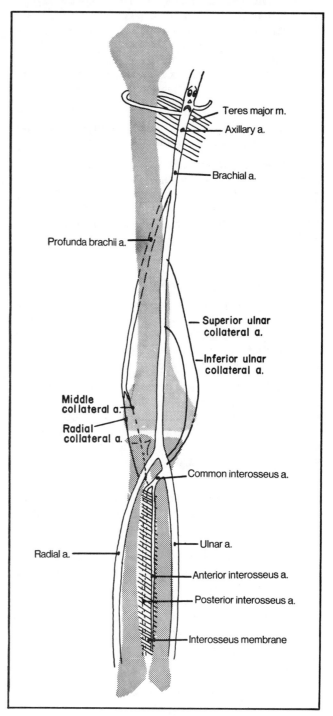

Figure 6-27

Fig. 6-28. Arteries of the hand. Look at the palm of your hand. One very prominent skin crease (U) goes to the ulnar aspect of the hand. The superficial palmar arch lies approximately at the level of this crease and is supplied mainly by the ulnar artery. A more proximal skin crease (R) extends to the radial side of the hand. The radial artery contributes to the deep palmar arch which lies about one thumb breadth proximal to this crease. Branches from the superficial and deep palmar arches combine to vascularize the fingers.

The palmar carpal arch is approximately located at the crease of the wrist, just proximal to the carpal bones and is largely fed by the anterior interosseus artery.

The dorsal carpal arch (fig. 6-28) is the only significant arch on the dorsum of the hand. It receives anastomoses from all the major arteries of the wrist. The dorsal carpal arch also contributes branches to the fingers. As a result of anastomoses, it is plain that if one severs any of these arches, the surgeon must tie off both ends of the cut to stop bleeding.

Arteries Of the Thorax

The internal thoracic artery, which originates from the subclavian artery and runs near the sternum, supplies part of the circulation to the intercostal spaces. Most of the circulation, however, comes from intercostal arteries that originate from the descending aorta.

Fig. 6-29. The intercostal arteries. The intercostal arteries anastomose with branches of the internal thoracic artery. Intercostal spaces 1 and 2 are supplied by branches of the subclavian artery as the aorta does not extend that high. The "intercostal" artery below rib 12 is called the subcostal artery as it obviously cannot be called an intercostal artery.

Fig. 6-30. The relationship between the intercostal arteries and the intercostal muscle layers (examine numbers 1-12 in the figure in sequence). The intercostal arteries have a choice as to whether to go between the two inner or the two outermost intercostal muscle layers. The two most external intercostal muscles point in opposite directions (fig. 4-28) and the intercostal arteries find this very confusing. Therefore, the arteries travel instead between the **internal** and **innermost intercostal muscles**, which run more or less in the same direction.

Note that the intercostal arteries also give off branches to the spinal cord.

Other branches of the aorta include:

a. coronary arteries - described with the heart (fig. 6-5)
b. bronchial arteries. These provide oxygenation to the bronchi and lungs. The pulmonary arteries cannot do this as they carry unoxygenated blood. The pulmonary alveoli do not have a bronchial artery circulation as they receive

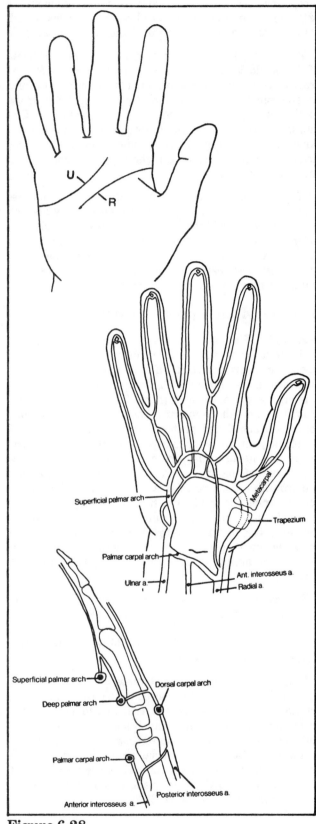

Figure 6-28

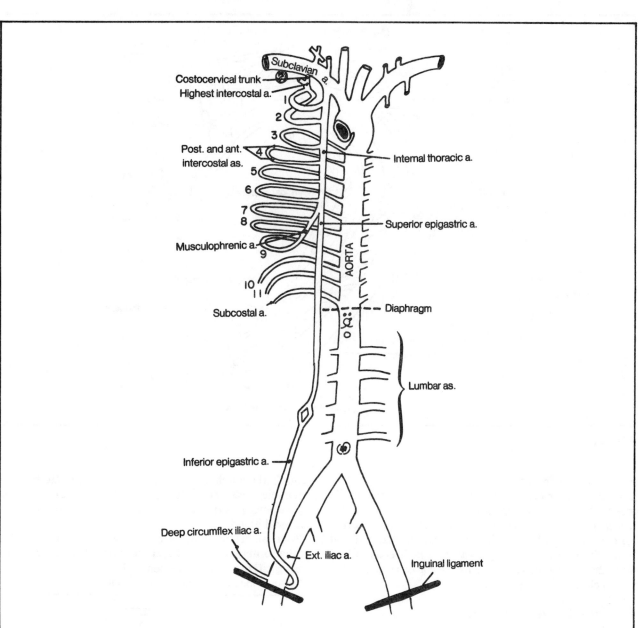

Figure 6-29

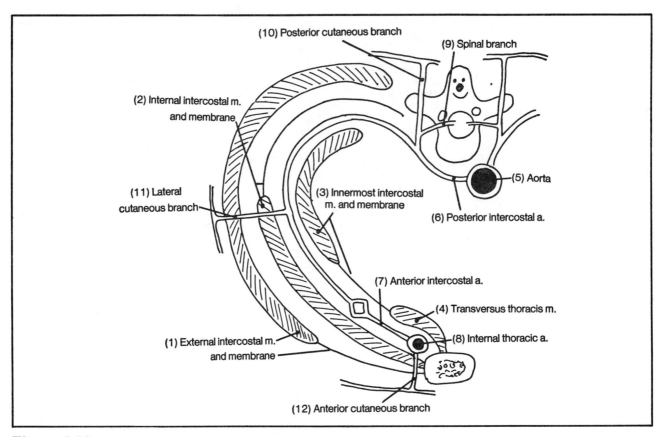

Figure 6-30

oxygen directly from the air.

c. esophageal arteries. The esophagus and bronchi lie anterior to the descending aorta.

d. arteries to the parietal pericardium. The internal thoracic artery also supplies branches to the parietal pericardium. The visceral pericardium is directly adherent to the heart and is supplied by the coronary arteries.

e. branches to the diaphragm

Arteries To the Abdomen

The descending aorta hugs the thoracic and lumbar vertebrae. On entering the abdomen, via the aortic opening of the diaphragm, the descending aorta becomes a scarecrow hung up against a pole (the lumbar vertebrae)(fig. 6-31). The scarecrow has eyes, a nose bent to the left, a moustache, a mouth, arms, breasts, belly button and legs. The umbilicus of the scarecrow does not underlie the human umbilicus but is somewhat higher. The human umbilicus lies at about the level of the scarecrow's pubic region.

Fig. 6-31. The abominable abdominal scarecrow.

(1) inferior phrenic arteries (scarecrow's eyes) - supply the diaphragm from below. Superior phrenic arteries supply the diaphragm from above and originate from the thoracic aorta. The inferior phrenic artery also gives off **a superior suprarenal artery**, which supplies the adrenal gland.

(2) celiac trunk (nose) - gives rise to the **hepatic and splenic arteries** (right and left sides of moustache) and **left gastric artery** (nose bent to left)

(3) superior mesenteric artery (mouth)

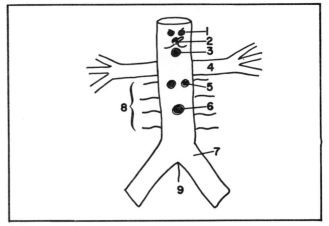

Figure 6-31

(4) renal arteries (scarecrow's arms)

(5) ovarian (or testicular) arteries (scarecrow's breasts)

(6) inferior mesenteric artery (scarecrow's umbilicus)

(7) common iliac arteries (scarecrow's legs)

(8) 4 paired lumbar arteries (scarecrow has a hairy trunk). These would be intercostal arteries except that there are no longer any ribs in this region. The lumbar arteries supply the abdominal wall, and also spinal cord, via branches that enter the intervertebral foramina.

(9) median sacral artery (scarecrow is male). This artery gives off a fifth pair of lumbar arteries.

The lumbar arteries, like the intercostal arteries, run around the trunk. They run between the internal oblique and transversus abdominis muscles (the homologues of the internal and innermost intercostal muscles of the thorax, between which the intercostal arteries run). The anterior abdominal wall is vascularized not only by the lumbar arteries, but, near its midline in the rectus region, is also supplied by the superior epigastric artery (via the internal thoracic artery) and inferior epigastric artery (from the external iliac artery - see fig. 6-29).

Fig. 6-32. Arterial supply of the GI tract. In this schematic figure the intestines are stretched out to show the vascular supply of the celiac trunk, and superior and inferior mesenteric arteries.

Fig. 6-33. The celiac trunk. The abominable scarecrow's nose has grown like Pinocchio's. His moustache has also grown and has become frayed on the right.

(1) left gastric artery - supplies the stomach and lower portion of the esophagus

(2) splenic artery - runs along the superior border of the pancreas. It supplies the spleen, but also gives off branches to the stomach and pancreas.

(3) hepatic artery (frayed side of moustache) - supplies the liver but also gives branches to the stomach and duodenum and pancreas via the right gastric(3A) and gastroduodenal (4) arteries

(4) gastroduodenal artery - supplies stomach via right and left gastroepiploic arteries(4A and 4B) and duodenum and pancreas via the superior pancreaticoduodenal artery(4C). Note that **the pancreaticoduodenal artery, of all the celiac artery branches, supplies the farthest point down the GI tract (the second portion of the duodenum). The superior mesenteric vessels then take over distribution beyond this point** (see fig. 6-32). Note that the superior pancreaticoduodenal artery anastomoses with the superior mesenteric artery system via the inferior pancreaticoduodenal artery.

(5) right hepatic artery - goes to the right lobe of the liver

(6) left hepatic artery - goes to the left lobe of the liver

(7) cystic artery - supplies the gall bladder. It usually is a branch of the right hepatic artery but has important anatomic variations in its origin and course (fig. 9-10). The vasculature in this area must be identified clearly before ligating the cystic artery in gall bladder surgery. Otherwise, the surgeon may ligate the right hepatic artery by mistake, causing death of the right side of the liver and possibly the patient!

Note in figure 6-33 (also fig. 9-6) that if the superior mesenteric artery were a saw that cut anteriorly, it would cut through the pancreas and stomach. If it were to cut posteriorly, it would cut through the pancreas and duodenum, right down to the inferior mesenteric artery.

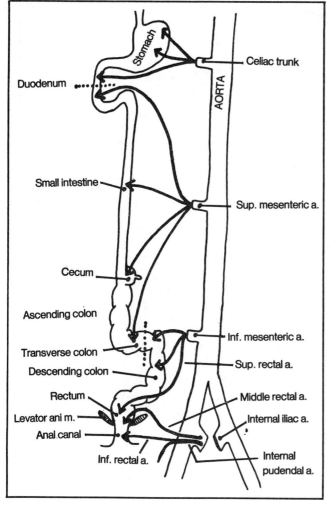

Figure 6-32

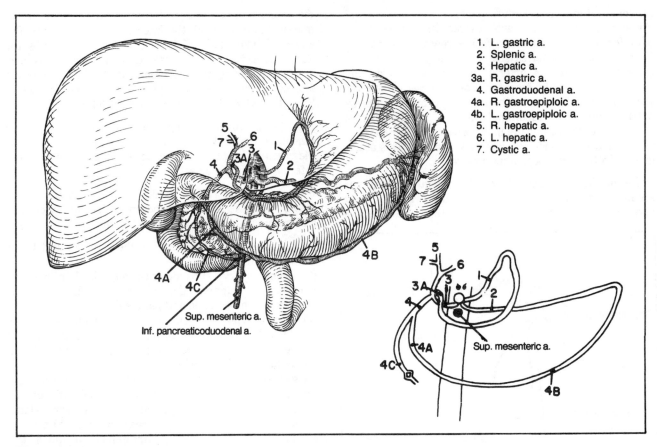

1. L. gastric a.
2. Splenic a.
3. Hepatic a.
3a. R. gastric a.
4. Gastroduodenal a.
4a. R. gastroepiploic a.
4b. L. gastroepiploic a.
5. R. hepatic a.
6. L. hepatic a.
7. Cystic a.

Sup. mesenteric a.
Inf. pancreaticoduodenal a.

Sup. mesenteric a.

Figure 6-33

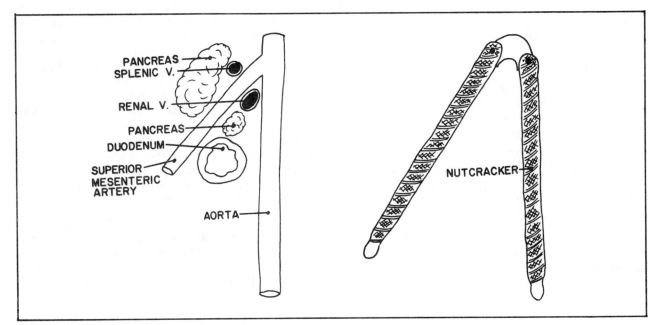

PANCREAS
SPLENIC V.
RENAL V.
PANCREAS
DUODENUM
SUPERIOR
MESENTERIC
ARTERY
AORTA

NUTCRACKER

Figure 6-33A

Fig. 6-33A. The superior mesenteric artery nutcracker. The superior mesenteric artery angles acutely with the aorta, like a nutcracker. The duodenum, pancreas(that portion of it called the **uncinate process**), and the left renal vein are the nuts. If the angulation is particularly acute, the duodenum may be compressed, causing symptoms of intestinal obstruction. If the left renal vein is compressed, there may be signs of renal failure.

It may be difficult to surgically remove clots from the origin of the superior mesenteric artery, as this zone is blocked anteriorly by the pancreas and splenic vein (fig. 6-33A).

Pancreatic disease may cause thrombosis and enlargement of the splenic vein, leading to an enlarged spleen.

Fig. 6-34. The superior and inferior mesenteric arteries. In a sense the **superior mesenteric artery** and **marginal artery (M)** may be considered to be a long vascular channel that extends along the length of the small and large intestine, the superior mesenteric artery extending along the small intestine and the marginal artery along the large intestine to the sigmoid colon. The intestines fold over on one another and various cross-linkages occur between the various parts of the large arterial channels. These include **ileocolic, right colic,** and **middle colic arteries** from the superior mesenteric artery and **left colic** and **sigmoid branches** from the inferior mesenteric artery.

The inferior mesenteric artery continues inferiorly as the **superior rectal artery** (fig. 6-34). The superior rectal artery supplies the rectum and part of the anal canal. It is joined by anastomoses from the **middle and inferior rectal arteries** which arise directly or indirectly from the internal iliac artery (fig. 6-32).

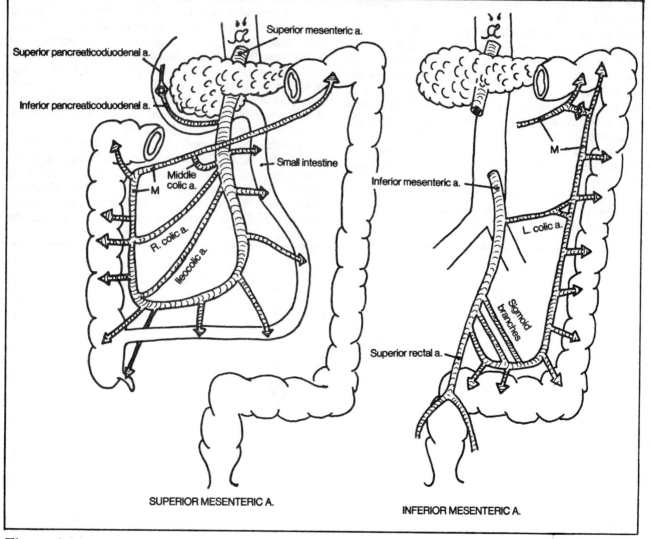

Superior mesenteric a.

Superior pancreaticoduodenal a.

Inferior pancreaticoduodenal a.

Small intestine

Middle colic a.

M

R. colic a.

Ileocolic a.

M

Inferior mesenteric a.

L. colic a.

Sigmoid branches

Superior rectal a.

SUPERIOR MESENTERIC A.

INFERIOR MESENTERIC A.

Figure 6-34

Arteries supplying the pelvis include:

a. superior rectal a. - a branch of the inferior mesenteric a. (fig. 6-34)
b. ovarian as. (fig. 6-31)
c. median sacral a. (fig. 6-31)
d. branches of the internal iliac a. (fig. 6-35)

Iliac Arteries

Fig. 6-35. Branches of the internal iliac artery. The internal iliac artery has numerous and variable branches. It supplies the pelvis, perineum, and buttocks. Actually, what happens is that this artery arises in the pelvis (the pelvis being that area above the levator ani muscle), recognizes the complexity of this region and tries to escape through whatever route is available. The numbers below refer to the numbers in figure 6-35.

(1) iliolumbar artery - tries to escape by climbing the pelvic wall; supplies iliacus, psoas, and quadratus lumborum muscles.
(2) superior and inferior gluteal arteries. These arteries supply the gluteal (buttock) region. They find the way blocked by the piriformis muscle which stretches from the sacrum to femur and fills the greater sciatic foramen, as the piriformis, too, is anxious to leave the region (fig. 4-45). The gluteal arteries squeeze by the piriformis, with the superior gluteal artery traveling above and the inferior gluteal artery traveling below this muscle.
(3) obturator artery - finds a nice hole, the obturator foramen, through which it exits to supply the thigh and hip.
(4) internal pudendal artery - performs a neat stunt. It enters the perineum by going out the greater sciatic foramen and back through the lesser sciatic foramen to get under the levator ani muscle (fig. 4-33,34). Once in the perineum it can give off its branches to the anal canal (inferior rectal artery) and external genitalia. The internal pudendal artery is the "artery of erection". Thus, occlusion of the internal iliac artery may cause impotence, which may be relieved by removing the obstruction.
(5) lateral sacral arteries - enter and hide within the sacral canal, supplying the meninges and nerve roots
(6) external iliac artery - leaps out of the pelvis under the inguinal ligament to become the femoral artery.

Other arteries unfortunately find no escape route and must remain in the pelvis. These include:

(7A) superior vesical arteries - supply the bladder, which lies totally within the pelvis.
(7B) inferior vesical artery - supplies the region inferior to the bladder, i.e., the seminal vesicles and prostate. Now, the ductus deferens (fig. 11-7) happens to extend from the prostate to the testicle. Realizing this, the inferior vesical artery in a flash of inspiration finds a novel escape route. It gives off an **artery to the ductus deferens** that follows the latter into the scrotum, to anastomose there with the testicular artery (fig. 6-37).
(7C) uterine artery. The uterine artery does not have the option of exiting to a testicle. It does, however, send branches to the vagina, which lies in the perineum; it also anastomoses with the ovarian artery which originates from the descending aorta (fig. 6-36).
(7D) middle rectal artery - supplies the rectum and part of the anal canal. Recall that the rectum lies totally in the pelvis, i.e., above the levator ani muscle (the anal canal lies in the perineum). The middle rectal artery anastomoses with the superior rectal artery (the continuation of the inferior mesenteric artery) as well as with the

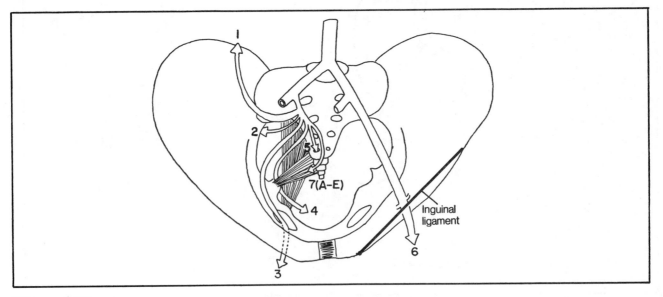

Figure 6-35

inferior rectal artery (a branch of the internal pudendal artery in the perineum)(fig. 6-32).

(7E) umbilical arteries. Arteries by definition are vessels that carry blood away from the individual's heart, whether in the fetus or in the adult. In the fetus, umbilical arteries therefore carry blood away from the fetus to the mother via the umbilical cord. There are two umbilical arteries, as they originate from the two internal iliac arteries (figs. 9-11 and 9-12). In the adult, the termination of the umbilical arteries is obliterated, although the artery still gives off other branches to the pelvis, commonly the superior vesical arteries, mentioned above.

Fig. 6-36. Anastomoses of the uterine and ovarian arteries.

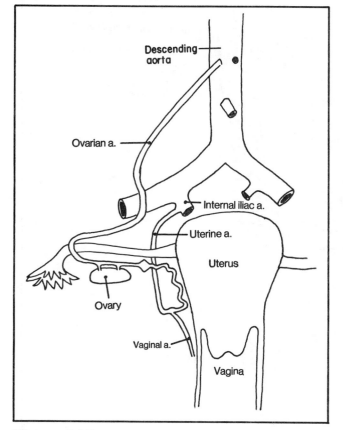

Figure 6-36

Fig. 6-37. Anastomoses of the testicular artery. Blood vessels in the spermatic cord include the **testicular artery** (from descending aorta), **artery of the ductus deferens** (arising indirectly from the internal iliac artery), and the **cremasteric artery** (arising indirectly from the external iliac artery).

The external iliac artery's main function is to pass under the inguinal ligament and become the femoral artery, which supplies the lower extremities. It has little time for branching in the pelvis. It does, however, produce two relatively small branches just prior to reaching the inguinal ligament, as follows (fig. 6-37):

(1) deep circumflex iliac artery - runs parallel to the inguinal ligament to supply the lower lateral wall of the abdomen

(2) The inferior epigastric artery - mentioned previously in the context of its connecting indirectly with the neck, via the umbilicus. It supplies the lower medial aspect of the abdomen. It arises near the inguinal ligament and is a landmark in distinguishing the two types of inguinal hernia. **Direct inguinal hernias arise medial to the inferior epigastric artery; indirect inguinal hernias lie lateral** (see also fig. 11-5).

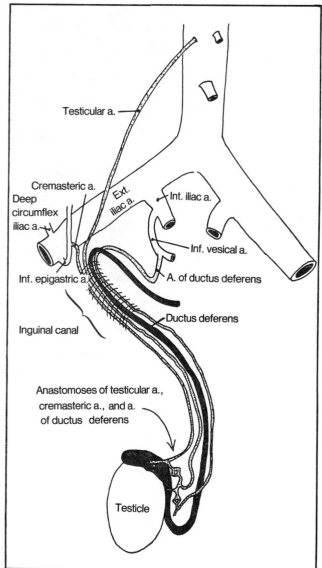

Figure 6-37

Arteries Of the Lower Extremities

The arterial supply to the lower extremities formally begins at the inguinal ligament where the external iliac artery becomes the femoral artery.

In large degree the lower extremity circulation parallels the pattern of the upper extremity. Below the level of the middle third of the femur, however, the lower extremity differs from the upper extremity in that the lower extremity appears to have been rotated 180 degrees. Thus, the knee (corresponds to the elbow) lies anteriorly and the plantar surface of the foot (corresponds to the palm) lies posteriorly (when on tiptoes). The arteries similarly rotate. Comparisons of figures 6-38 and 6-27 are useful.

Fig. 6-38. Arteries of the right lower extremity.

(1) The femoral artery corresponds to the brachial artery of the upper extremity.

(2) The profunda femoris artery corresponds to the

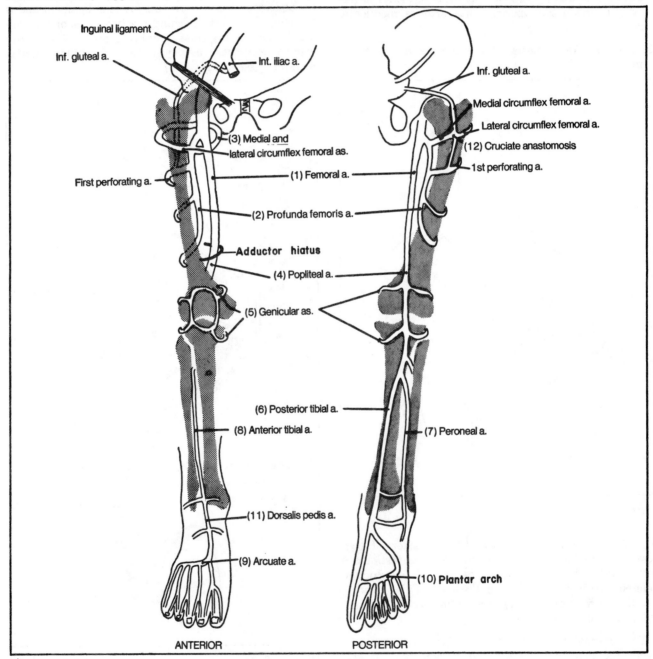

Inguinal ligament
Inf. gluteal a.
Int. iliac a.
Inf. gluteal a.
Medial circumflex femoral a.
Lateral circumflex femoral a.
(3) Medial and lateral circumflex femoral as.
(12) Cruciate anastomosis
First perforating a.
(1) Femoral a.
1st perforating a.
(2) Profunda femoris a.
Adductor hiatus
(4) Popliteal a.
(5) Genicular as.
(6) Posterior tibial a.
(8) Anterior tibial a.
(7) Peroneal a.
(11) Dorsalis pedis a.
(9) Arcuate a.
(10) Plantar arch
ANTERIOR
POSTERIOR

Figure 6-38

profunda brachii. It is the main artery supplying the thigh.

(3) Medial and lateral circumflex femoral arteries correpond to anterior and posterior humeral circumflex arteries.

(4) The popliteal artery resembles the brachial artery where the latter crosses the elbow . The femoral artery becomes the popliteal artery at the hiatus of the adductor magnus muscle. Occlusion of the femoral artery commonly occurs at this point.

(5) The genicular arteries correspond to anastomoses around the elbow.

(6) The posterior tibial artery corresponds to the radial artery. In fact, clinicians commonly feel for the posterior tibial pulse near the medial malleolus just as they feel for the radial pulse near the distal radius.

(7) The peroneal artery corresponds to the ulnar artery.

(8) The anterior tibial artery corresponds to the posterior interosseus artery. There does not appear to be a homologue of the anterior interosseus artery of the forearm.

(9) the arcuate artery corresponds to the dorsal carpal arch.

(10) The plantar arch corresponds to the superficial and deep palmar arches.

(11) The dorsalis pedis artery is a continuation of the anterior tibial artery. It is an important artery for palpation of the pulse in the foot.

(12) The cruciate anastomosis of the thigh - consists of connections between the **medial and lateral circumflex femoral arteries** and branches from the **inferior gluteal artery** and **first perforating artery**. This important anastomosis provides a way to bypass a femoral artery occlusion by an indirect connection with the internal iliac artery.

The Venous Circulation

Much of the anatomy of the venous system parallels that of the arterial system. Venous anatomy is much more variable than arterial anatomy. Consequently, it does not pay to memorize it in too much detail. In this section the differences between the venous and arterial anatomy will be emphasized.

There is a major difference between the overall layout of the venous and arterial circulations. The aorta is a single major vessel which extends branches along its course to the head and neck, upper extremities, thorax, abdomen, and lower extremities. The venous circulation, in contrast, consists of four main divisions:

A. The superior vena cava and its divisions
B. The inferior vena cava and its divisions
C. The azygous system
D. The portal circulation

Fig. 6-39. The superior vena cava and its divisions.

(1) superior vena cava
(2) brachiocephalic v.
(3) subclavian v.
(4) internal jugular v.
(5) external jugular v. Note the "W"-shaped structure where the external jugular vein crosses the internal jugular vein, arising from the junction of facial(6), retromandibular(7) and posterior auricular(8) veins.
(9) maxillary v.
(10) anterior jugular v. - connects with the base of the external jugular vein. The left and right anterior jugular veins communicate with each other across the midline, as the **jugular venous arch**(not shown), a point of caution to be noted in avoiding hemorrhage during a tracheostomy procedure.
(11) median cubital v. of arm - a common site for blood drawing

There is no venous correspondence to the aortic arch or descending aorta. Instead the superior vena cava, which supplies the upper part of the body, divides into two brachiocephalic veins which become subclavian, axillary, brachial, and radial and ulnar veins, as in the arterial system. Many of the side branches of these veins correspond to and follow their respective arteries of similar name. For instance, a plexus of vertebral veins follows the vertebral artery and ends in the subclavian vein.

The internal jugular vein is the homologue of the carotid artery. It really corresponds to both the internal and external carotid arteries because it not only subserves the brain (as does the internal carotid artery), but also connects with the face (as does the external carotid artery). Figure 6-39 shows how the facial vein connects directly with the internal jugular vein. Other veins of the head may connect either with the internal or external jugular veins. There is a rich anastomotic network. The internal jugular vein is the only one of the jugular veins to directly drain the brain. The others are involved with more superficial tissues but may indirectly connect with the brain through **diploic veins**, which extend through the skull to the dura.

What happens to the axillary vein as it reaches the upper extremity is as simple as ABB'C (see fig. 6-39). The Axillary vein (A) divides into a Brachial (B) and Cephalic (C) vein. The brachial vein (actually a pair of veins - **venae comitantes**) travels deep, next to the brachial artery. The cephalic vein travels **superficially**. The brachial vein itself gives rise to the Basilic (B') vein, another superficial vein (fig. 6-39).

A glance at your upper extremity will tell you that veins, rather than arteries, tend to lie close to the skin.

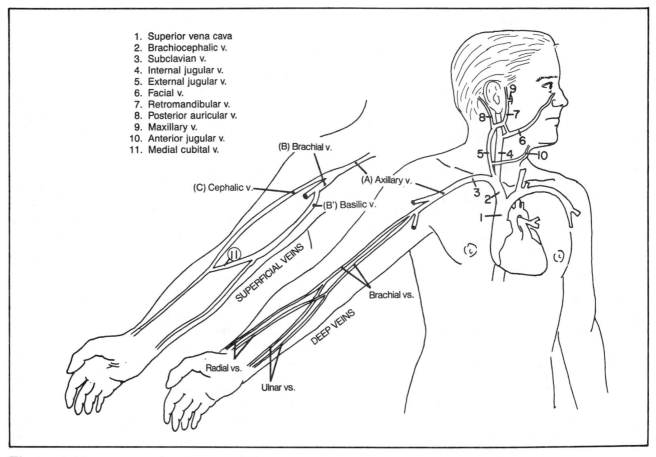

1. Superior vena cava
2. Brachiocephalic v.
3. Subclavian v.
4. Internal jugular v.
5. External jugular v.
6. Facial v.
7. Retromandibular v.
8. Posterior auricular v.
9. Maxillary v.
10. Anterior jugular v.
11. Medial cubital v.

(B) Brachial v.

(C) Cephalic v.

(A) Axillary v.

(B') Basilic v.

SUPERFICIAL VEINS

Brachial vs.

DEEP VEINS

Radial vs.

Ulnar vs.

Figure 6-39

These veins are tributaries of the cephalic and basilic veins. The term cephalic means "head" or "main". The cephalic vein extends along the wrist and it is the "main" vein for extracting blood at the wrist.

Venous Drainage Of the Brain (figs. 6-40 and 6-41)

Fig. 6-40. Venous drainage of the brain. Unlike other arteries of the body which have corresponding veins, Willis (fig. 6-21) has no female counterpart. This is because he is so ugly that the veins flee in the opposite direction, jumping clear out of the brain and directly into the **dural sinuses.** You see, the brain is separated from the cranial bone by a PAD (P-pia, A-arachnoid, D-dura membranes - fig. 6-18), otherwise know as the **meninges.** The **pia** is thin and vascular (Willis lives in it) and hugs the brain. The **arachnoid** lies between the pia and dura, is arachnoid-like, like a gossamer spider web, and avascular. The **dura** lies up against the bone and is thick and durable, containing a double layer of connective tissue (the outer layer representing periosteum) with thick venous channels, called **sinuses,** that lie between the two layers. The dura dips down between the cerebral

hemispheres as the **falx cerebri** and between the cerebrum and cerebellum as the **tentorium cerebelli.** The "PAD" surrounds the entire central nervous system, including the spinal cord and optic nerve.

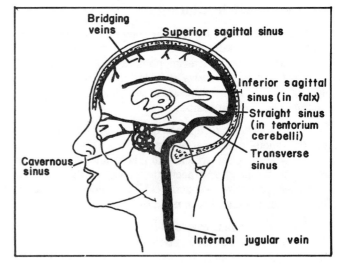

Figure 6-40

Three main venous sinuses deserve special mention (fig. 6-40):

1) Spinal fluid drains into the **superior sagittal sinus.**

2) The **cavernous sinus**, into which venous blood drains from the eye, provides a potential source of entry into the brain of orbital and facial infections. Many important structures run by or through the orbit. These can be damaged by disease in this critical area. Sometimes the carotid artery breaks open into the cavernous sinus (a "fistula") resulting in blood backing up into the orbital veins and a protruding eye with dilated blood vessels.

3) The **transverse sinus** runs by the ear and may become involved in inner ear infections.

Fig. 6-41. Venous drainage at the base of the skull.

(1) ophthalmic v. - drains into the cavernous sinus
(2) sphenoparietal sinus
(3) cavernous sinus
(4) superior petrosal sinus - runs along the petrous portion of the temporal bone
(5) inferior petrosal sinus
(6) sigmoid sinus - a continuation of the transverse sinus to the internal jugular vein
(7) transverse sinus

The cranial venous sinuses connect with the vertebral plexus of veins that is associated with the spinal cord dura mater. It is believed that tumor cells originating as far away as the prostate may spread to the brain along this plexus.

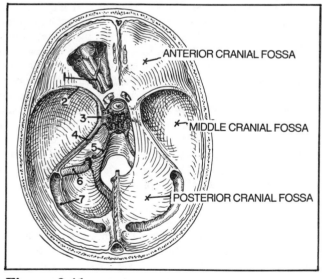

ANTERIOR CRANIAL FOSSA

MIDDLE CRANIAL FOSSA

POSTERIOR CRANIAL FOSSA

Figure 6-41

The Azygous, Inferior Vena Caval, and Portal Venous Systems

There is no single vein that corresponds to the descending aorta. The inferior vena cava extends from the right atrium through the diaphragm to the abdomen. The problem with this arrangement is that the thoracic part of the inferior vena cava is so short; there is little room for it to give off intercostal vessels, as the descending aorta does. This is the job of the azygous system, which gives off intercostal veins (fig. 6-42).

Fig. 6-42. The azygous, inferior vena caval, and portal venous systems.

The Azygous System

In analogy with the arterial system, the first intercostal spaces do not drain into the azygous system but, somewhat higher, into the brachiocephalic veins. The internal thoracic vein (not show in fig. 6-42) follows the internal thoracic artery; it has anterior intercostal branches just as does the internal thoracic artery. These veins anastomose with the intercostal branches of the azygous vein. Apart from intercostal branches, the azygous vein drains the lungs, esophagus, pericardium, vertebrae, diaphragm and thoracic cord (not shown in fig. 6-42).

Once beyond the rib cage, the azygous vein commonly joins the renal vein and becomes the **ascending lumbar vein** of the abdomen . The ascending lumbar vein gives off lumbar veins, which supply the abdominal wall, traveling in the same corresponding muscles layers as in the thorax (between internal oblique and transversus muscles). **The intercostal and lumbar veins correspond to the intercostal and lumbar arteries.**

The azygous system thus connects with the superior vena cava as well as with the tributaries of the inferior vena cava. These anastomoses are significant as they provide alternate routes for blood to get back to the heart, if the inferior vena cava becomes obstructed.

The Inferior Vena Cava (fig. 6-42)

Within the abdomen, the branches of the inferior vena cava resemble those of the aorta with the following exceptions:

1. The left testicular vein comes off the left renal vein rather than from the inferior vena cava (recall that both the right and left testicular arteries come off the descending aorta). **Varicocele (venous dilation) in the left side of the scrotum of an older person should make one suspect blockage of the left testicular vein by tumor growing along the left renal vein.**

2. The inferior vena cava does not drain the digestive tract. Note that in the arterial circulation (fig. 6-31) there are three main unpaired arterial trunks extending from the midline of the abdominal aorta - the **celiac trunk** (supplies stomach, liver, and spleen), the **superior mesenteric artery** (supplies small intestine and part of

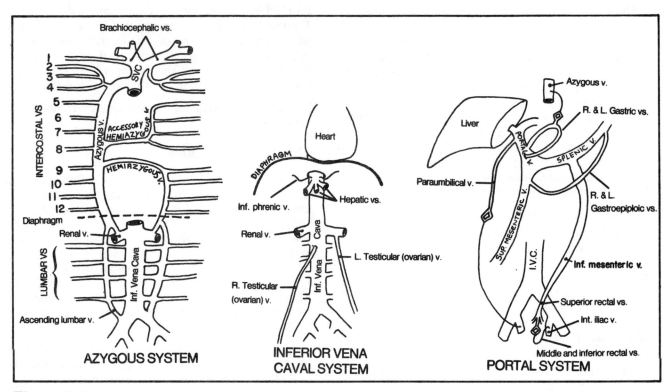

Figure 6-42

large intestine) and the **inferior mesenteric artery** (supplies the last part of large intestine, including the rectum). **The venous system, however, appears as if someone cut off all these gastrointestinal trunks (except for the hepatic veins, which drain the liver) and fused them into a single separate system: the PORTAL SYSTEM.**

The Portal Venous System (fig. 6-42)

Blood from the far reaches of the intestinal tract enters the liver, carrying absorbed food from the digestive system. Blood then exits the liver via the hepatic veins to enter the inferior vena cava at the level of the diaphragm (fig. 9-12).

Fortunately, there are other ways for the portal circulation to reach the heart than through the liver, for sometimes, e.g., in cirrhosis, the liver is so damaged that sufficient portal blood cannot pass through it. Under such circumstances, alternate circulatory routes may develop:

1. The gastric veins may communicate indirectly with the azygous vein through varices in the lower esophagus (fig. 6-42). Such varices may result in upper gastrointestinal hemorrhage.

2. The superior rectal vein may communicate through varices (hemorrhoids) with the middle and inferior rectal veins, which arise from the internal iliac vein. Such hemorrhoids may protrude and/or bleed.

3. Paraumbilical veins that normally drain into the portal vein from the umbilicus may communicate indirectly with the iliac veins. Varices then form around the umbilicus (**caput medusae**). Note that the paraumbilical veins are not the same as the umbilical vein. The latter brings blood from mother to fetus and is obliterated after birth, becoming the **round ligament (ligamentum teres)**. The paraumbilical veins extend along the round ligament in the adult and simply are those veins that drain blood from the abdominal wall around the umbilicus. (See the fetal circulation in fig. 9-11). Tumors may sometimes spread from the liver along the paraumbilical veins and present as a mass at the umbilicus.

Generally, abdominal veins **above** the umbilicus drain toward the superior vena cava (for instance, through the intercostal veins). Veins **below** the umbilicus drain toward the inferior vena cava (e.g., through the lumbar veins). There is extensive anastomosing among these groups of veins; veins over both the abdominal and thoracic walls may become dilated with obstruction of either the superior or inferior vena cava. By compressing the superficial veins and noting the direction of blood flow, one can distinguish between a superior and an inferior vena caval obstruction.

Veins Of the Pelvis and Perineum

Like the arteries of the pelvis, the veins are highly variable in their courses, so much so that they are referred to as a plexus, around the bladder, prostate (or uterus), rectum and vagina. In general, these veins drain into the internal iliac veins. However, these vessels also connect with the superior rectal vein, which drains into the inferior mesenteric vein, thus providing anastomoses between the portal and inferior vena caval venous systems.

Recall that the azygous system communicates with the superior and inferior vena cava; the portal system communicates with the azygous system at the level of the esophagus and the inferior vena caval system at the level of the internal iliac artery. Thus, the **inferior vena caval, azygous, and portal systems all intercommunicate.**

Veins Of the Lower Extremities

As in the upper extremity, the veins in the lower extremity are of two types - superficial and deep. The deep ones generally follow the arterial system and have similar names.

Fig. 6-43. Deep and superficial veins of the lower extremity. The deep veins include:

(1) external iliac v.
(2) femoral v.
(3) profunda femoris v.
(4) popliteal v.
(5) knee vs.
(6) peroneal v.
(7) anterior tibial v.
(8) posterior tibial v.

As in the upper extremities (compare with fig. 6-39) there are 2 main superficial veins (great and small saphenous veins). The **great saphenous** is the larger one, arising high, where the extremity begins (comparable to the cephalic vein of the upper extremity). It runs down to the big toe, just as the cephalic vein runs down to the thumb. The **small saphenous** vein arises lower down, near the back of the knee from the popliteal vein, and corresponds to the basilic vein, although the basilic vein arises quite high (fig. 6-39). The superficial and deep veins of the lower extremity communicate with one another in a rich anastomotic network. Poor venous return may lead to varicose veins.

Note that the great saphenous vein runs just anterior to the medial malleolus. This a favored area for a venous "cutdown" in the lower extremity, when it becomes necessary to find a vein for the delivery of intravenous fluids.

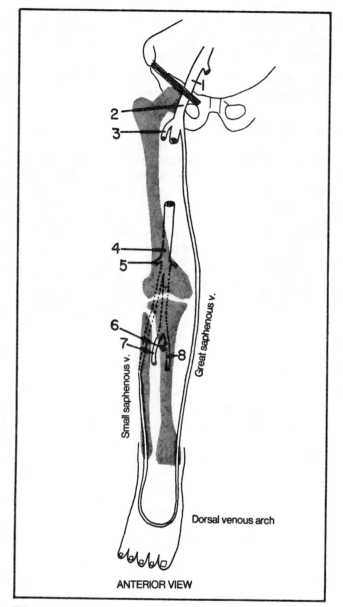

Figure 6-43

Lymphatics

The intercellular spaces of the body contain lymph fluid. This fluid drains along lymph vessels to eventually reach the venous circulation in the neck. Entry into the veins occurs at the point where the jugular and subclavian veins become the brachiocephalic vein (fig. 6-45). There are variations.

Lymph does not drain from the brain and spinal cord because the central nervous system contains no lymphatics. The central nervous system is more than content to have its own special cerebrospinal fluid which drains into the superior sagittal sinus (fig. 12-4).

Lymph fluid filters through lymph nodes, which lie along the lymph channels, kill microorganisms and produce lymphocytes and antibodies. There are two main components to the lymph circulation:

1. A superficial system that drains the skin (fig. 6-44).
2. A deep system that drains deeper structures (fig. 6-45).

Fig. 6-44. The superficial component of the lymph system - drainage of the skin. Drainage of the skin goes to three main groups of lymph nodes: **axillary nodes, cervical nodes,** and **inguinal nodes.** These are the main groups of nodes that one palpates on routine physical exam. The continental divide of the Western United States is a line along the Rocky Mountains that separates two major river drainage systems. Rivers to the west of the divide enter into the Pacific ocean whereas rivers to the east enter the Atlantic. Two such divides (dotted lines in fig. 6-44) surround the body, one at the level of the **umbilicus** and one at the level of the **clavicle.**

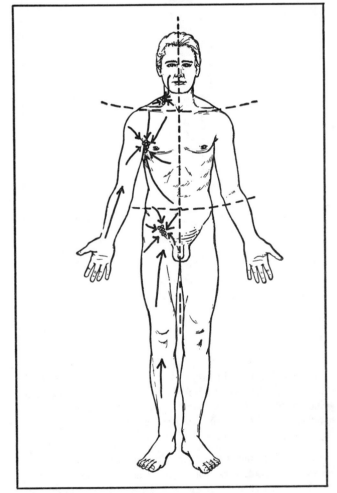

Figure 6-44

Lymphatics draining the skin above the clavicle extend to the cervical glands. Those between clavicle and umbilicus enter the axillary glands. Those below the umbilicus enter the inguinal glands. A malignant melanoma of the skin that is unfortunately situated at the level of the umbilicus may spread to lymph nodes in both axillae and both inguinal areas.

In addition to receiving lymph from the total body skin, the axillary, inguinal, and cervical lymph nodes also receive lymph from deeper tissues of the **upper and lower extremities, head,** and **neck.** Lymph flow then continues to the deep lymph system.

The deeper structures of the thorax, abdomen, pelvis and perineum tend to drain directly into the deeper lymph system, rather that passing to the superficial axillary, inguinal, and cervical nodes.

Fig. 6-45. The deep lymph system. The axillary, cervical and inguinal nodes drain into the subclavian, jugular, and right and left lumbar lymphatic trunks respectively.

The **cysterna chyli** is an expansion at the lower end of the thoracic duct, at the level of vertebra L2.

The **bronchomediastinal trunks** are especially important in that they drain the deep thoracic viscera, most notably the lung. Lymph from the lung tends to drain to nodes at the **hilum,** or root, of the lung, which lies in the vicinity of the **carina** (the region of bifurcation of the trachea into right and left branches). From there the lymph enters the bronchomediastinal trunks. The right lung and lower left lung tend to drain to the right bronchomediastinal trunk. Lymphatics of the upper left lung tend to drain along the left bronchomediastinal trunk. Thus, tumor within nodes of the right bronchomediastinal trunk may have arisen from the left lower lung.

Deep in the body, lymph channels lie in and around all the major organs. In general, the lymph channels with their associated nodes drain along the major arteries and veins (e.g., iliac or mesenteric vessels) toward one central trunk, either the **right lymphatic duct** or the **thoracic duct.** The thoracic duct enters only the left venous circulation. Most of the body lymph, in fact, empties on the left side. Only the right head and neck, right upper extremity and right upper trunk empty on the right, through the right lymphatic duct. In the abdomen, pelvis and perineum, all the deep lymphatics eventually enter the thoracic duct. In the thorax, some lymph channels enter the thoracic duct directly. Other channels extend directly upward as the right and left bronchomediastinal trunks. For instance, mediastinal nodes drain the lungs, tracheobronchial tree, heart, and esophagus via the bronchomediastinal trunks.

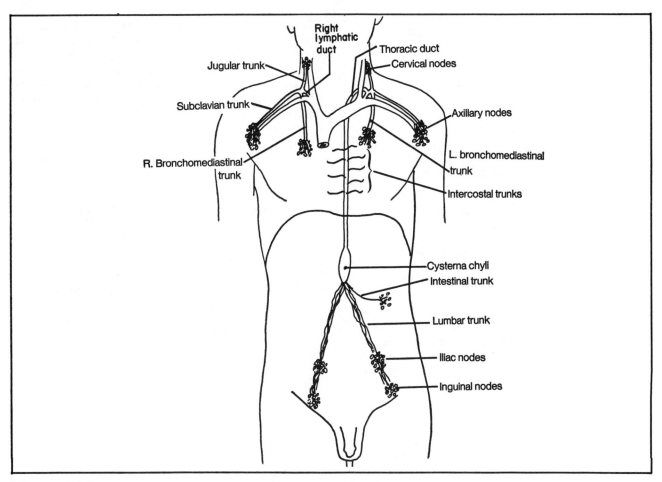

Figure 6-45

Since the lymphatic system eventually reaches the venous system in the supraclavicular area, the presence of supraclavicular nodes (especially on the left side) should raise the suspicion of possible metastatic spread of tumor from the thorax or abdomen.

Another important lymph pathway is the **internal thoracic (internal mammary) chain of nodes** which follows the internal thoracic blood vessels and eventually enters the brachiocephalic vein directly or joins lymphatics in the vicinity of the brachiocephalic vein.

As examples of the difference between the superficial and deep circulations, a (lymphatic- spreading) tumor of the scrotal skin tends to spread to the inguinal nodes, which may be readily palpable. A tumor of the testes, however, tends to spread more deeply to the iliac nodes and the latter may not be palpable. Anal tumors above the pectinate line (fig. 4-41) tend to spread to the deeper nodes, whereas tumors below the pectinate line spread to the groin.

Lymph glands on either side of the diaphragm may communicate in either direction with one another. Hence, malignancies of the thorax may spread to the abdomen and vice-versa. Tumors of the esophagus may spread to the stomach or the reverse may occur; this consideration becomes important in determining how much to remove surgically when an area is affected by tumor.

Fig. 6-46. Lymph drainage of the breast. Superficial skin drainage goes to the axillary nodes. Deeper breast tissue generally also drains to the axillary nodes, when a tumor is on the lateral aspect of the breast (most breast malignancies occur in the upper lateral quadrant of the breast). However, tumors in the upper inner quadrant of the breast may spread to the internal thoracic chain. Tumors in the lower inner quadrant of the breast may spread via lymphatics to the umbilicus, which itself connects deeply with lymphatics surrounding the liver. There is normally a certain degree of anastomosing between the various lymphatic systems and, in conditions of tumor blockage of the main channels, alternate channels may be used, with tumor spreading to the other breast or to cervical glands.

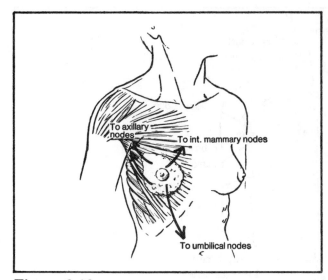

Figure 6-46

The **thymus gland** (fig. 10-1) is composed largely of lymphoid tissue. It is especially prominent in childhood. It has an important role in the immune system in the development and release of T lymph cells. It may have other important endocrine functions.

Fig. 6-47. The spleen (shaded area), lies within the rib cage. It usually underlies ribs 9-11 and is hard to palpate, except when enlarged. It functions to filter the blood of assorted debris, to store blood, and, as a lymphoid organ, to contribute to the production of lymphocytes and plasma cells. Sometimes, there are extra ("accessory") spleens, which are important to identify and remove in those operations where removal of the spleen is important.

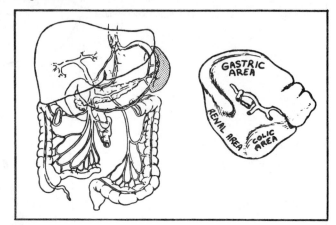

Figure 6-47

CHAPTER 7. THE SKIN AND FASCIA

Skin looks roughly similar from one area to the other. Yet dermatology is replete with classic disorders that affect select areas of the body. The rash in lupus often affects the bridge of the nose. In secondary syphilis the rash may be prominent on the palms and soles. Chicken pox affects the trunk more so than the extremities. Sometimes the reasons for a particular distribution are clear. Some conditions may localize to moist regions that contain many sweat glands. Herpes zoster (shingles) presents in a strip-like distribution, because it affects the nerve along a single dermatome (fig. 12-8). In other conditions the basis for a given disorder is not clear. The skin may contain hidden anatomical maps as yet undiscovered.

Fig. 7-1. The map of skin cleavage lines, reflecting parallel bundles of collagen fibers within the skin. Surgical scars heal with less scarring when the incision is made parallel to these lines. The lines may be invisible but frequently run parallel to skin creases. They tend to run longitudinally within the extremities and circumferentially in the neck and trunk. The map-like pattern of skin **lymph** drainage, however, does not necessarily follow cleavage lines (fig. 6-44). Cutting parallel to a cleavage line may in some cases cut perpendicular to lymph channels, leading to swelling from blocking of lymph drainage. E.g. an extensive incision below the lower eyelid might produce a cosmetically acceptable scar but cause lid swelling from lymph obstruction.

More questionable kinds of maps, e.g. acupuncture maps, have been used in asian countries for many centuries for the diagnosis and treatment of various disorders. The belief is that stimuli at select points may affect the functioning of distant organs. There is no clear observable anatomic basis for such maps. It presently is controversial as to whether success in acupuncture relates to specific points or to non-specific factors, such as suggestion, or non-specific release in the brain of natural pain-reducing substances (endorphins).

Fig. 7-2. Layers of the SCALP spell "SCALP". The layer of loose connective tissue is a "danger" area for hemorrhage or infections, which may spread throughout its extent. This layer provides the plane in which a scalping occurs. Hemorrhage or edema in the loose connective tissue layer area may spread far anteriorly to cause puffy eyes, as the frontalis muscle does not connect directly to bone. Hemorrhage between the pericranium (periosteum) and bone (for example, **cephalhematoma** occurring during childbirth) generally does not spread beyond the sutures of the underlying bone, because the pericranium adheres to the suture line. The dense connective tissue between the aponeurotic layer and skin also resists the spread of hemorrhage.

In injecting the scalp with local anesthesia for a laceration, the injection should meet resistance. If it doesn't, it has been placed too deeply, in the loose connective tissue layer, where it will be ineffective.

It is important to suture the aponeurotic layer when it is severed, to assure tight wound closure.

Fig. 7-3. The breast. The milk-secreting tissue of the breast consists of about 15-20 **lobes**. Each lobe may be likened to a bunch of grapes converging on stems (**lactiferous ducts**) that end in the nipple. A dilated portion (the **lactiferous sinus**, or **ampulla**) is present before the exit point of the lactiferous ducts. The **areola** is the dark circular zone surrounding the nipple. It contains **areolar glands**, which produce a lubricating secretion that facilitates nursing. Fibrous **septa** (**suspensory ligaments of Cooper**) separate the lobules, attach to the skin, and lend support to the breast. The softness

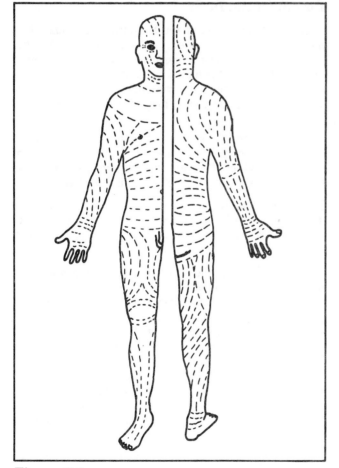

Figure 7-1

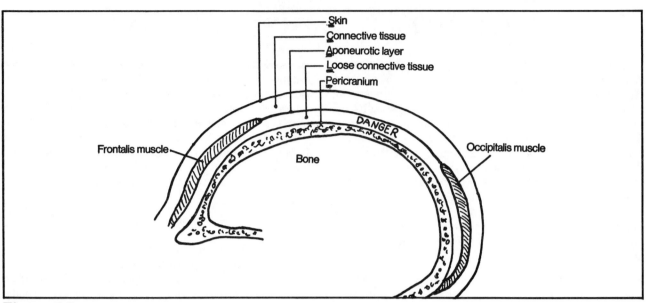

Figure 7-2

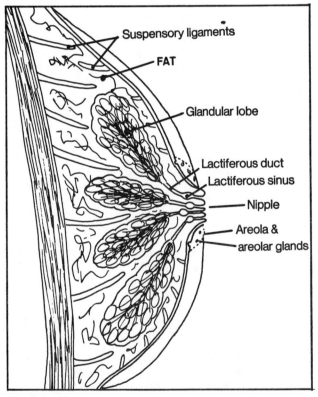

Figure 7-3

of a breast is due to fat. The normal nodular feel reflects the glandular lobes. Malignant tumors involving the fibrous septa may cause dimpling of the attached skin. Malignant tumors of the lactiferous ducts may induce nipple retraction. In palpating a breast it is important to extend the exam into the axilla, as part of the glandular tissue of the breast reaches the axilla.

For lymph drainage of the breast, see figure 6-46.

Fascia Of the Body

The skin and fascia of the body may be likened to a house, in which the outside paint coat represents the skin, the shingles represent superficial fascia, and the outer and inner walls of the house, which interconnect with one another, are the deep fascia with its connecting partitions. The contents of the rooms are the various organs and body spaces.

The presence of fascial compartments is important clinically in that they may locally restrict the spread of infection or tumor. They are important to the surgeon as they form important planes of surgical dissection.

Fig. 7-4. Fascial space in the neck. An abscess developing posterior to the prevertebral fascia may break through into the retropharyngeal space or spread laterally.

Fig. 7-5. Fascia of the abdominal wall. As in other body areas, the abdominal wall contains a superficial and deep layer of fascia. The deep layer is simply a thin layer that covers the external oblique muscle. The superficial fascia is quite prominent, especially in fat people, and, unlike other body areas, may itself be divided into two separate layers - **superficial** and **deep**. The superficial layer of the superficial fascia (**fatty layer; Camper's fascia**) contains fat. The deep layer of the superficial fascia (**membranous layer; Scarpa's fascia**) contains fibrous tissue which may provide support for sutures during wound closure of the abdomen.

The anatomy of the fascial layers of the abdomen resembles that of a skin diver wearing a total body suit and, underneath, only the front 1/2 of a pair of underpants (fig. 7-5). The body suit is Camper's (fatty) fascia which is continuous with the superficial fascia of the rest of the body (but is especially fatty in the abdomen). The 1/2 pair of underpants is Scarpa's (membranous) fascia which is present only on the anterior abdomen and perineum; it fades out laterally and superiorly and is not present posteriorly. The skin of the diver is the deep fascia which is continous with deep fascia in other body areas.

If a patient experiences a ruptured urethra, urine may collect deep to Scarpa's (membranous) fascia, just as urine will collect deep to one's underpants in a situation of incontinence. The urine will not enter the thigh, however, as the underpants are tight-fitting (Scarpa's fascia is fused to the deep fascia of the thigh). The above analogy is not perfect, however, as Scarpa's (membranous) fascia does not completely surround the penis but extends only up to the glans penis. In the perineum, Scarpa's fascia is called **Colle's fascia**.

The deep fascia of the thigh (**fascia lata**) is tough. It resembles a support hose stocking surrounding the thigh muscles. The lateral aspect of the fascia lata is particularly tough and is called the **iliotibial tract**; the latter stretches from the iliac crest to the tibia and fuses

with the tendons of tensor fascia lata and gluteus maximus. Surgeons sometimes use fascia lata for grafting purposes.

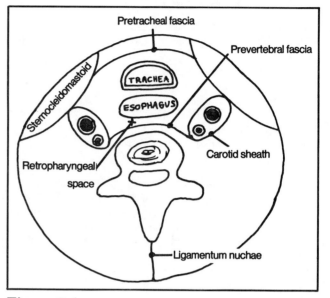

Figure 7-4

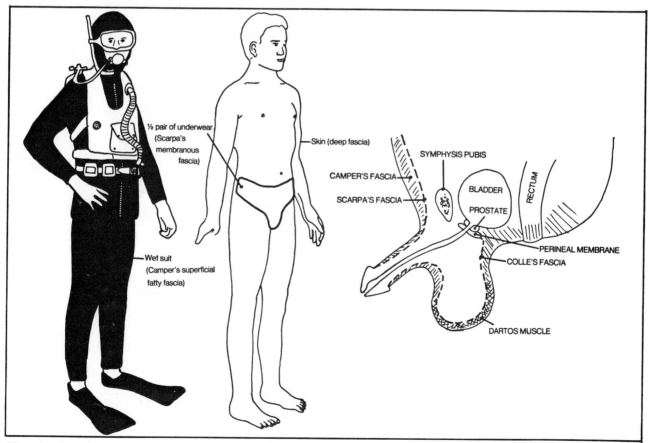

Figure 7-5

CHAPTER 8. THE RESPIRATORY SYSTEM

Fig. 8-1. The nasal pasages.

(A) opening of the auditory (Eustachian) tube
(E) openings of anterior, middle, and posterior ethmoid sinuses
(F) opening of frontal sinus
(Fr) frontal sinus
(HS) hiatus semilunaris
(M) opening of maxillary sinus
(N) opening of nasolacrimal duct
(P) pharyngeal tonsil (adenoids)
(S) sphenoid sinus

A spelunker crawling into a nose would find that the medial wall of the nasal passageway is relatively smooth - the **nasal septum**. The lateral wall, however, contains a series of ridges - the superior, middle and inferior nasal **conchae**. Below the ridges are recesses called the superior, middle and inferior **meatuses**. There is one more meatus - the one above the superior conch, called the **sphenoethmoidal recess** (because it is at the junction of the sphenoid and ethmoid bones).

The explorer would not stay long in the inferior meatus because it not uncommonly rains there; the inferior meatus contains the opening of the **nasolacrimal (tear) duct**. If the explorer were to climb up into the middle meatus (to avoid the flood) he would see quite an astonishing groove - the **hiatus semilunaris**. Looking into this groove he would see the entrance into a vast dark cavern - the **maxillary sinus**. Also entering into or near this groove are:

1. a long connection from the frontal sinus.
2. several entry points of ethmoid sinus cells. On contemplating the anatomy of the hiatus semilunaris , it would become clear to the explorer why so many people suffer from lingering sinusitis; the exit from the maxillary sinus is high up on the maxillary sinus wall, making drainage by gravity difficult, unless the patient's head is tilted. Moreover, inflammatory drainage from a frontal or ethmoidal sinus may enter into the maxillary sinus, thus compounding the matter.

Since the ethmoid bone contains both superior and middle conchae, but not an inferior conch, which is a separate bone in itself, the ethmoidal air cells (or sinuses) are only associated with the superior and middle conchae, exiting below them.

Climbing higher, the explorer might be bored with the superior meatus, because it only contains one or two more ethmoidal air cell openings. Still climbing, the explorer would not be bored with the sphenoethmoidal recess, because, in addition to ethmoidal air cell openings, it

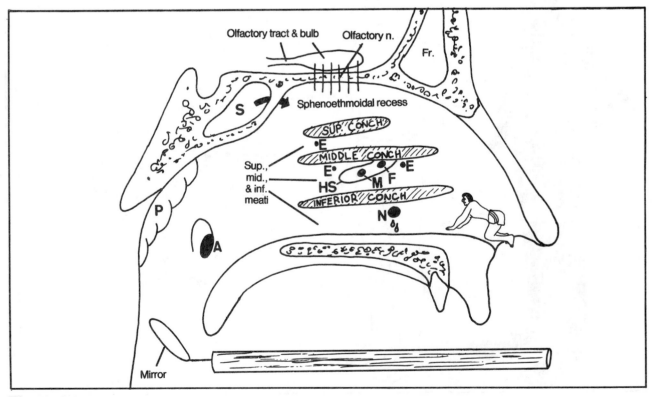

Figure 8-1

contains the opening into the **sphenoidal sinus**. In addition, the sphenoethmoidal recess contains the olfactory mucosa with its associated **olfactory nerve fibers**.

Fig. 8-2. View through the mirror in figure 8-1.

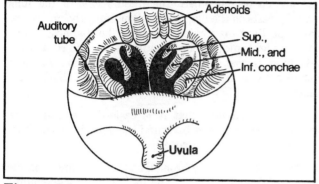

The Lungs

Like the heart, each lung is surrounded by visceral and parietal membranes (the visceral and parietal pleura). This is analogous to two fists pushing into 2 balloons (see figs. 6-8 and 8-3).

Fig. 8-3. The pleura. Normally, there is only a thin film of fluid between the visceral and parietal pleural layers. When fluid collects, as in inflammation (**pleurisy**) or tumor effusions, it characteristically collects by gravity in the **costodiaphragmatic recess** where it can be extracted by syringe.

Fig. 8-4. The technique for extracting fluid from the pleural cavity. The level of fluid can usually be detected on x-ray or by noting areas of dullness upon percussion of the back over the lungs. Usually the 8th intercostal space (between ribs 8 and 9) is chosen. Normally the

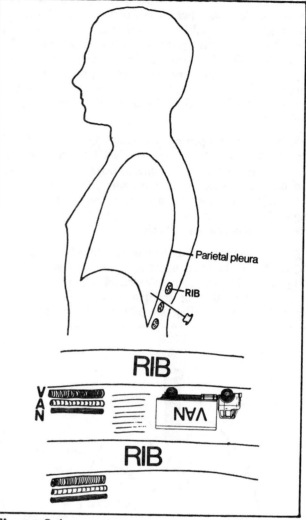

Figure 8-4

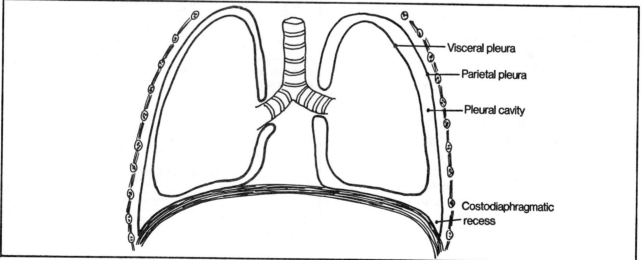

Visceral pleura

Parietal pleura

Pleural cavity

Costodiaphragmatic recess

inferior edge of the scapula lies over rib 7, or over the 7th intercostal space (which lies between ribs 7 and 8). The needle is inserted into the 8th intercostal space just **over** rib 9 rather than just under rib 8, in order to avoid hitting the "VAN" (intercostal Vein, Artery, Nerve) that runs under each rib. Care must be taken not to puncture the lung, as this may lead to leakage of air between the layers of pleura and collapse of the lung (**pneumothorax**).

Fig. 8-5. Gross topography of the lung and associated bronchi. For orientation, the trachea bifurcates at the sternal angle of Louis (the junction between manubrium and body of the sternum). The right primary bronchus is slightly wider and has a slightly steeper downward angle than the left. As a result, foreign bodies more frequently lodge on the right side.

As the heart is localized more on the left side of the thorax, the right lung, upset about its lack of a heart, compensates by acquiring an extra lobe. There are three lobes to the right lung and two to the left, each lobe containing corresponding secondary and tertiary lobar bronchi. The bronchi continue to subdivide, becoming microscopic bronchioles and, ultimately, alveoli, where gaseous exchange occurs with the blood.

The three lobes of the right lung are not simply stacked one on top of the other. The middle lobe of the lung actually rests on the diaphragm. Consequently, an abscess from the liver may extend through the diaphragm and cause a right middle lobe pneumonia. The **lingula** of the **left** lung is that part of the left upper lobe which is homologous with the right middle lobe and touches the diaphragm.

In figure 8-5 note that it is easiest to examine the inferior lobes from the back and the superior lobes from the front, using a stethoscope.

There are 10 tertiary bronchi on the right and 9 on the left. Each is associated with an individual bronchopulmonary segment (fig. 8-6).

Sometimes there is an accessory, or **azygous** lobe at the lung apex on the right side, where the azygous vein indents the superior lobe.

Fig. 8-6. Schematic structure of the lung bronchopulmonary segment. Connective tissue septa separate the individual segments. Structures entering the segment (pulmonary a., bronchial a., tertiary bronchus and its subdivisions) tend to lie centrally; structures leaving the lung (pulmonary veins, lymphatics) lie near the septa. Disease processes such as infection, tumor, and collapse of a portion of lung (**atelectasis**) commonly remain confined to individual bronchopulmonary segments. The thoracic surgeon must know the locations of the various bronchopulmonary segments as the clinical condition may require the removal of isolated segments without compromising the surrounding lung tissue.

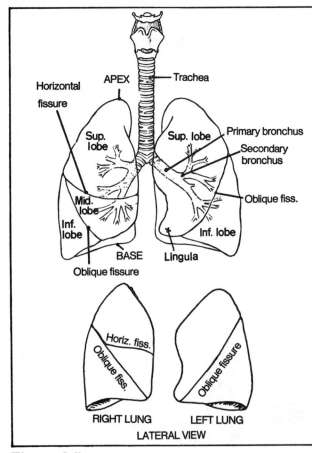

Figure 8-5

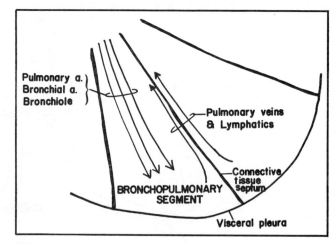

Figure 8-6

CHAPTER 9.   THE DIGESTIVE SYSTEM

The Tongue

For muscles of the tongue and swallowing, see FIGURES 4-76 THROUGH 4-83.

Fig. 9-1.   Papillae of the tongue. These contain chemoreceptive taste buds that relay taste information to the brain through cranial nerve 7 (anterior 2/3 of the tongue) and cranial nerve 9 (posterior 1/3 of the tongue). If you have ever been licked by a giraffe, you will have noted that the filiform papillae provide a somewhat rough surface (like a FILE), that assists in digesting food by abrading it.

The VALlate papillae lie in a "V"-shaped arrangement on the back of the tongue (where you might place a VALium tablet prior to swallowing). These large papillae contain a particularly great concentration of taste buds.

Fungiform papillae lie on the tip and sides of the tongue, the areas that are exposed to air on slight protrusion of the tongue (leaving the tongue locally exposed to outside FUNGI).

The **foramen cecum** is the original opening of the embryonic thyroglossal duct, which connected thyroid

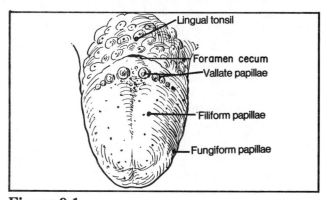

Figure 9-1

gland to tongue. Normally absent in the adult, this duct sometimes persists and contains a **thyroglossal duct cyst**, or ectopic thyroid tissue. These conditions may present as swellings in the posterior 1/3 of the tongue.

Fig. 9-2.   The salivary glands. The parotid gland hides behind the ramus of the mandible. The **parotid duct (Stenson's duct)** reaches laterally around the masseter

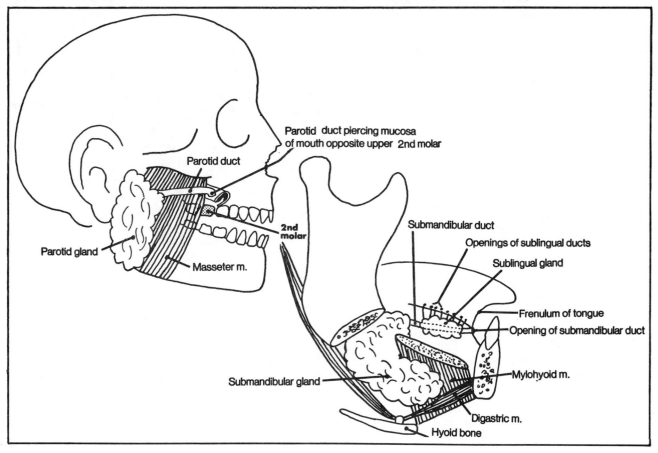

Figure 9-2

muscle to pierce the mucosa of the inner cheek, entering the mouth opposite the upper second molar. Note the location of the parotid gland. **Mumps**, which induces parotid gland inflammation, causes swelling above the angle of the mandible.

The submandibular (submaxillary) gland hides behind the mylohyoid muscle, partly lying on either side of it. Its submandibular duct enters the mouth at the base of the frenulum of the tongue (figs. 9-2, 9-3).

The sublingual gland lies lateral to the submandibular duct. It issues multiple sublingual ducts that enter the mouth through multiple openings along the lateral base of the tongue (figs. 9-2, 9-3).

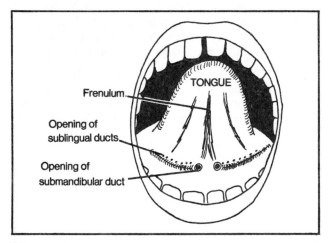

Figure 9-3

Fig. 9-4. The digestive tract. The upper 1/4 of the esophagus contains skeletal (voluntary) muscle, whereas the lower 3/4 contains smooth muscle. Thus, the upper 1/4 is supplied by somatic motor fibers of cranial nerve 10 (branches of the recurrent laryngeal nerve, which also innervates the larynx and pharynx - fig. 14-14). The lower 3/4 of the esophagus is supplied by parasympathetic nerve fibers of cranial nerve 10 that extend as the esophageal plexus to the stomach (fig. 14-14). The jejunum begins where the duodenum ends. The upper 2/5 of the small intestine is the **jejunum**, the lower 3/5 the **ileum**. Together they measure about 20 feet. The ileum ends at the large intestine, specifically that portion called **the cecum**.

The esophagus contains three narrowed areas where foreign bodies may lodge. Tumors of the esophagus and chemical burns also have a predisposition to localize in these areas:

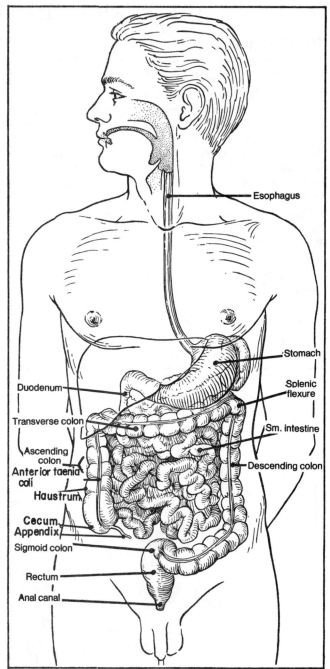

Figure 9-4

A. The junction between pharynx and esophagus
B. The level of the tracheal bifurcation (see fig. 17-12)
C. The junction between esophagus and stomach

The esophagus may also be narrowed by an enlarged left atrium. The left atrium is part of the posterior surface of the heart lying just below the tracheal bifurcation, in direct opposition to the esophagus. Such enlargement may be detected in a barium swallow x-ray.

Fig. 9-5. Hiatus hernia. In a **sliding hiatus hernia**, the esophagus and stomach are pulled up through the diaphragm. This form is the most common. When associated with an incompetent gastroesophageal sphincter, there commonly will be gastric acid backflow (reflux) and heartburn. In a **paraesophageal hiatus hernia**, the esophagus remains in place, but a portion of the stomach herniates. This form is generally not associated with reflux.

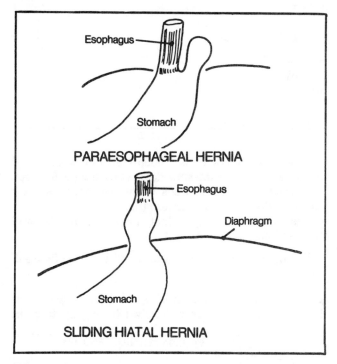

Esophagus

Stomach

PARAESOPHAGEAL HERNIA

Esophagus

Diaphragm

Stomach

SLIDING HIATAL HERNIA

Figure 9-5

Fig. 9-6. The relation between duodenum and pancreas. The pancreas resembles a fist that is holding onto 2 straws (superior mesenteric artery and vein) and punching the second portion of the duodenum. This pushes the second portion over to the right side of the abdomen. The third segment of duodenum crosses the midline while the fourth segment lies on the left. Note how the mesenteric vessels extend anterior to the third segment. Not shown is the **ligament of Treitz (suspensory ligament of the duodenum)**. This is a fibromuscular band which anchors the 4th segment of the duodenum to the right crus of the diaphragm.

The stomach overlies the pancreas. Tumors or ulcers of the stomach (which may cause epigastric pain) can erode into the pancreas, causing back pain.

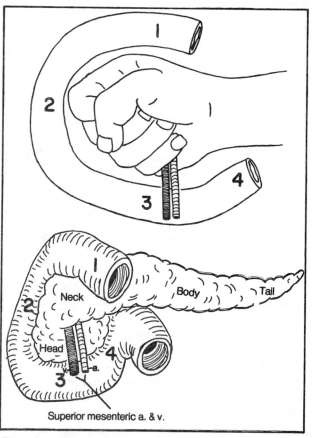

Neck Body Tail

Head

Superior mesenteric a. & v.

Figure 9-6

Fig. 9-7. The pancreatic and bile ducts. The pancreas is partly an **endocrine** organ, secreting hormones into the blood stream (insulin and glucagon), and partly an **exocrine** organ, secreting digestive enzymes that are carried by ducts directly to the duodenum.

Most commonly, the main pancreatic duct joins with the bile duct, to form the **ampulla of Vater** (hepatopancreatic ampulla), which enters the second part of the duodenum. Prior to this junction, the bile duct and main pancreatic ducts each contain a terminal sphincter muscle, which helps prevent bile and pancreatic enzymes from backing up along the wrong duct. With obstruction (e.g. by tumor) at the level of the ampulla of Vater, bile and enzymes cannot reach the duodenum, and pancreatitis and/or jaundice may result. The ampulla of Vater has its own sphincter (**sphincter of Oddi**) which, if spastic, may also result in obstruction. Pancreatitis may sometimes be averted by the presence of an **accessory pancreatic duct** which enters the duodenum as a smaller opening above the main opening. If the accessory duct connects with the main duct, then digestive enzymes may have an alternate route (the accessory duct) in the event of obstruction of the ampulla of Vater.

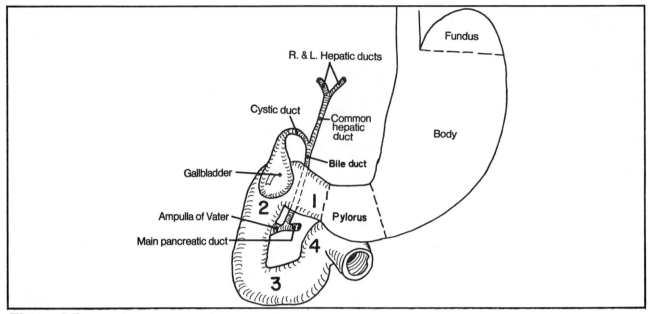

Figure 9-7

The bile duct passes behind the first section of duodenum and inserts (through the ampulla of Vater) into the second portion of duodenum. The gallbladder overlies the junction between the first and second portions of the duodenum and may be seen peeking out of the undersurface of the liver (fig. 9-8).

The large intestines may be distinguished from the small intestines by the following points. The large intestine:

a. is wider.
b. externally, has 3 longitudinal muscular bands called

teniae coli, and saccular bulges of the wall, called **haustra**. The small intestine does not have these features. The anterior tenia coli leads to the appendix, and is a useful landmark for the surgeon (fig. 9-4).
c. has fatty appendages that project from its wall **(appendices epiploica)**

For the anatomy of the rectum and anus see figure 4-41.

Fig. 9-8. The liver. Normally, the lower edge of the liver cannot be felt, or is barely felt on palpating the

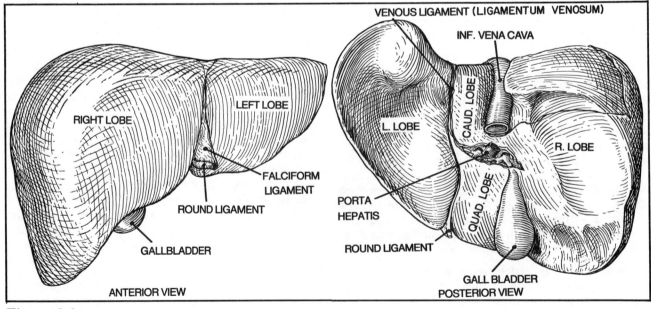

Figure 9-8

abdomen below the right rib margin. When enlarged, the liver may extend well below this point.

Fig. 9-9. Schematic view of the liver, anterior view. Imagine that the liver is transparent. A branding iron has impressed the letter "A" onto the posterior surface of the liver. The hand bar of the brand represents the hepatic duct, hepatic artery, and portal vein. Right and left branches of the latter structures enter the liver in the crossbar of the "A" (the **porta hepatis**). The "A" divides the liver into left, right, quadrate, and caudate lobes. The quadrate and caudate lobes anatomically appear as part of the right lobe, as they lie to the right of the falciform ligament. Functionally however, they belong to the left

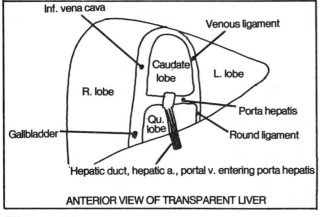

Figure 9-9

lobe, as they derive their blood supply from the left hepatic artery and left portal vein (and drain into the left hepatic duct). The leg (groove) of the "A" on the right side of the body contains the inferior vena cava above and the gall bladder below. The leg on the left is less important, containing the venous ligament and round ligament, which are remnants of the fetal circulation, reviewed in figures 9-11 and 9-12. The hepatic duct and artery, and portal vein do not settle in the "A" as they lie outside the liver in the lesser omentum (the peritoneal fold that connects liver with stomach (figs. 9-14 and 9-15).

Blood leaving the walls of the large and small intestine carry digested nutrients to the right and left branches of the portal vein. In addition, blood supplying oxygen to the liver enters by the right and left hepatic arteries. The need for such double circulation is also seen in the lung. The pulmonary arteries and bronchial arteries both supply the lungs, the pulmonary to obtain oxygen from the alveoli, the bronchial arteries to oxygenate lung tissue itself. The portal vein alone does not provide sufficient oxygenation. Interruption of the right or left hepatic artery may cause the corresponding lobe of the liver to die. Blood from the liver returns through hepatic veins to the vena cava.

Fig. 9-10. Arterial supply of the gallbladder. In most cases, the right hepatic artery lies sandwiched between the hepatic duct and portal vein (its a good thing it doesn't get pinched off), and gives rise to the cystic artery, which

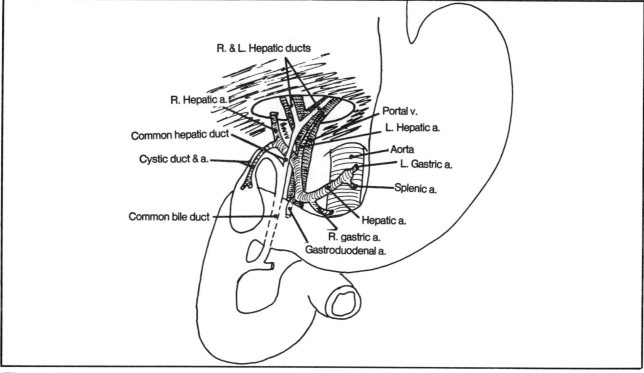

Figure 9-10

supplies the gall bladder. There are many variations, however. The right hepatic artery may cross in front of the hepatic duct or behind the portal vein. The hepatic artery may arise from the superior mesenteric artery, and the left hepatic artery may originate from the left gastric artery. The cystic artery may originate from right or left hepatic arteries. Therefore, on removing a gallbladder (**cholecystectomy**) it is important for the surgeon to identify the arterial connections with the gallbladder before tying off the cystic artery, to avoid inadvertently tying a hepatic artery, with consequent liver death.

Not shown in figure 9-10 is the inferior vena cava,

which lies posterior to the portal vein. These two veins sometimes are surgically anastomosed (**portocaval shunt procedure**) to divert portal blood to the vena cava, when liver circulation is obstructed, as in **cirrhosis**.

Fig. 9-11. Fetal circulation. Note how blood in the portal vein largely bypasses the liver, entering the vena cava directly through the ductus venosus. It shunts in this manner because prepared nutrients come from the mother and there is little need for the portal vein to filter intestinal blood through the liver. Similarly, blood bypasses the lungs by detouring through the foramen ovale and ductus arteriosus, as oxygenated blood is supplied by the mother.

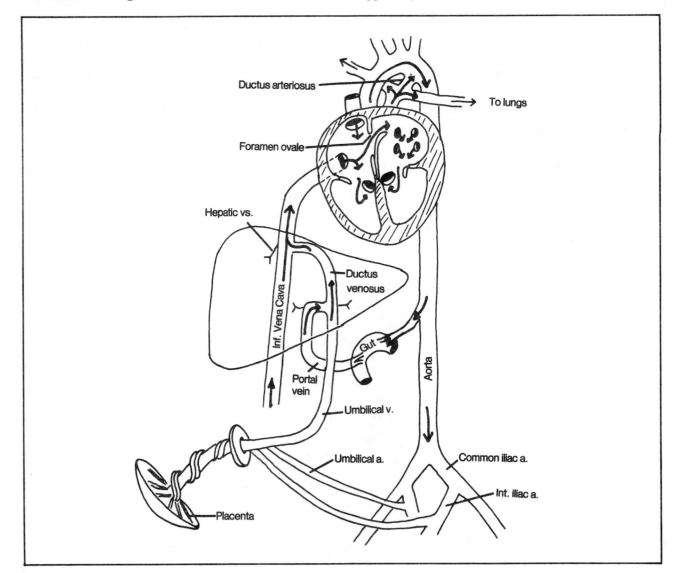

Figure 9-11

Fig. 9-12. Postnatal circulation. Major changes are closure of the foramen ovale and ductus arteriosus, and degeneration of the umbilical vein and ductus venosus to become the **round ligament (ligamentum teres)** and **venous ligament (ligamentum venosum)** respectively. There is also partial obliteration of the umbilical arteries. Compare with figure 9-9.

In some people the ileum contains a small outpouching, called **Meckel's diverticulum**. This is a remnant of the embryonic yolk stalk. It may contain misplaced (ectopic) gastric mucosa and be a source of gastrointestinal inflammation and hemorrhage.

Peritoneum

Hopefully, the following fictionalized story will help in understanding the complex peritoneal folds.

Once upon a time the abdomen had no peritoneum. A kindly surgeon opened the abdominal wall and saw the various organs lying squashed, one on top of the other (fig. 9-13).

"How cold these organs look", thought the surgeon, who then proceeded to cover them with a blanket (peritoneum) and tuck them in (figs. 9-13 through 9-16).

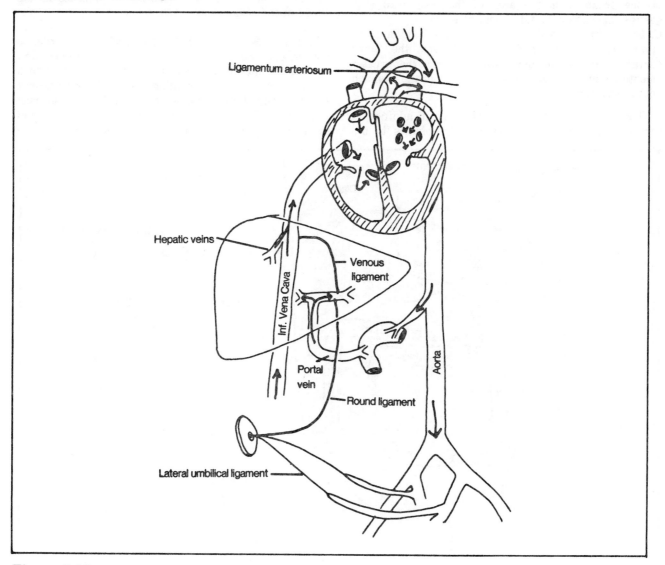

Figure 9-12

Fig. 9-13. Schematic view of the abdominal organs covered by a blanket (peritoneum, shaded area). The organs became tucked-in with the blanket. The object of the tuck-in was to enclose each organ with the blanket so that the organs no longer touched each other or the cold posterior abdominal wall. Double folds of blanket arose from these tuck-ins, and these folds were called by several names that in some senses were synonymous:

A. Mesentery - a fold specifically connecting an organ to the **posterior** abdominal wall (for instance, mesentery of small intestine; sigmoid mesocolon)

B. Ligament - a fold connecting one organ with another, or one organ with any area of the body wall (e.g. gastrosplenic ligament, falciform ligament)

C. Omentum - a fold connecting stomach to another organ (e.g., the lesser omentum connects the lesser curvature of the stomach with the liver; the greater omentum connects the greater curvature of the stomach with the transverse colon)

These folds proved of great value, for in addition to carrying blood vessels, nerves, and lymphatics to the various attached organs, they provided the organs with mobility.

The tuck-in process began with the liver. The liver, however, presented three difficulties:

1. The round ligament (remnant of the umbilical vein), which attaches the liver to the umbilicus, remained in the way, hanging in mid-air like a clothesline.

2. The bile duct, which connects the liver and second portion of duodenum, was also in the way.

3. The superior edge of the liver remained firmly adherent to the diaphragm and could not be detached by the encircling peritoneum. This region persisted as the **"bare area"** of the liver.

Fig. 9-14. Solution to the tuck-in problem. In order to circumvent the above difficulties, the surgeon first draped the mesentery over the round ligament (like a sheet over a clothesline), forming the **falciform ligament** fig. 9-14). In order to get around the bile duct, the surgeon left the blanket stretched taut over the bile duct and stomach and indented the blanket around the right side

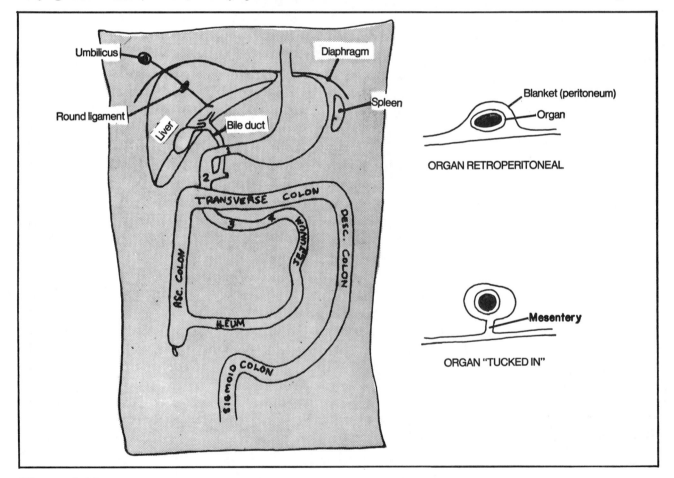

Figure 9-13

(gall bladder side) of the bile duct to form a pocket (**lesser sac** - arrow in fig. 9-14) behind the stomach. The double fold that resulted (**lesser omentum**) connected the stomach and proximal duodenum with the bile duct and liver. The opening into the lesser sac is the **epiploic foramen** (fig. 9-15). The distant end of the sac, which lay draping from the greater curvature of the stomach over the transverse colon (fig. 9-16), became the **greater omentum**. While on the posterior side of the bile duct, the surgeon spread the blanket around the remainder of the liver, except for the bare area of the liver, which stubbornly stuck to the diaphragm. The **coronary** and **triangular ligaments** (not shown) are the portions of peritoneum that form the border of the bare area.

It is important to remember that the bile duct, hepatic artery and portal vein all run together in the lesser omentum as part of the anterior wall of the epiploic foramen (figs. 9-10, 9-14, 9-15). A surgeon may thus control hemorrhage from a bleeding cystic artery (in gallbladder surgery, for instance) by placing the index finger in the epiploic foramen and the thumb anterior to the lesser omentum and then compressing the hepatic artery. This same maneuver is useful for palpating stones in the bile duct . For orientation, note that the epiploic foramen lies immediately to the left of the neck of the gallbladder.

The surgeon lifted up various other areas of the intestinal tract, surrounding them with peritoneum and pinching them off the posterior abdominal wall with the formation of mesenteries. These mesenteries are:

1. The transverse mesocolon, which connects transverse colon to posterior abdominal wall along a line that stretches from left kidney across the pancreas and second portion of the duodenum to the right kidney.
2. Mesentery of the small intestine which connects small intestine to posterior abdominal wall along a line that connects the duodenojejunal junction with the terminal ileum.
3. Sigmoid mesocolon.
4. Mesoappendix.

Also pinched off was the spleen, as is better shown in figure 9-15.

The remainder of the intestinal tract stayed retroperitoneal. Retroperitoneal structures include:

1. The entire duodenum (except for its first inch, which is referred to as the "duodenal cap")
2. The ascending and descending colon, and the rectum.
3. The pancreas, kidneys, ureters, adrenal glands, aorta and vena cava. Sometimes a kidney may have a mesentery ("floating kidney"). Torsion around such a

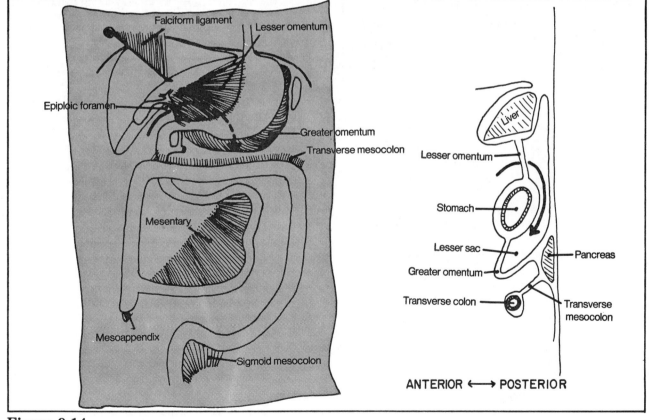

Figure 9-14

mesentery may sometimes compromise the circulation to the kidney.

The anal canal doesn't even enter into this discussion as it lies in the perineum, below the levator ani muscle, completely outside the abdominal cavity. The bladder, uterus, and ovaries do lie above the levator ani muscle and the peritoneum does cover them (figs. 11-8,11-11).

The fourth part of the duodenum looks very much like the jejunum. By definition, the duodenum becomes the jejunum when it becomes intraperitoneal (develops a mesentery). Along similar lines, the junction between sigmoid colon and rectum is that point where the sigmoid colon becomes retroperitoneal. The rectum, as you will recall, becomes the anal canal simply by passing below the levator ani muscle.

Fig. 9-15. Transverse section of the abdomen as seen from above. Note the relative positions of the hepatic artery, bile duct, and portal vein: "DUctus-DExter (to the right)" and "POrtal vein-POsterior".

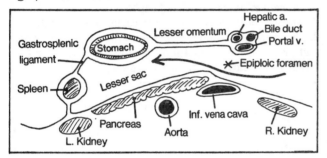

Figure 9-15

Fig. 9-16. The greater omentum. The greater omentum actually spreads out over the transverse colon and small intestine like an apron. In doing so, it fuses to the transverse colon and transverse mesocolon. The greater omentum is the first thing one sees on opening the abdominal cavity.

It is important to realize that the peritoneum, in addition to lining the abdominal organs and posterior abdominal wall, also lines the anterior abdominal wall. Rather than a blanket, the peritoneum is actually a balloon that is opened whenever the abdominal wall is opened. The inside of the balloon is the peritoneal cavity.

Normally the peritoneal cavity contains only a bare amount of fluid. Its walls are in direct apposition to one another. The walls slide against one another with minimal friction so as to provide great mobility to the viscera. Inflammation (**peritonitis**) may spread within the peritoneal cavity. Fluid may collect (**ascites**) in various pathologic conditions (e.g., cirrhosis of the liver) and air may enter, as during perforation of a gastric ulcer. Strange as it may seem, all the abdominal and pelvic organs lie outside the peritoneal cavity as the peritoneal cavity is the inside of the balloon.

The only normal opening into the peritoneal cavity is that of the fallopian (uterine) tubes. During ovulation the ovarian follicle ruptures into the peritoneal cavity and the egg makes a brief trip through the peritoneal cavity to the fallopian tubes.

Following abdominal surgery, healing may occur with adhesions, in which the inside walls of the peritoneum stick to one another. This may diminish motility or cause constriction of a portion of intestine.

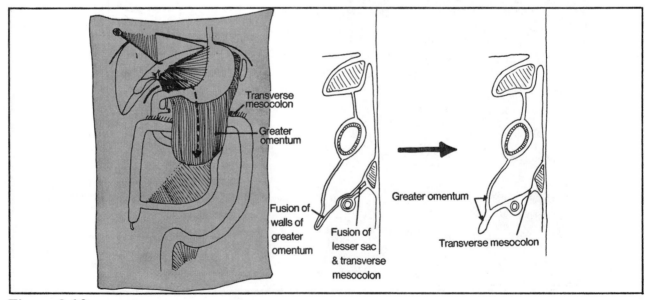

Figure 9-16

CHAPTER 10. THE ENDOCRINE SYSTEM

Fig. 10-1. Location of the major endocrine (ductless, hormone-secreting) glands of the body.

A. The pituitary gland - responsible for the release of hormones linked to metabolism, growth, sexual development and the immune response.

B. The pineal gland - appears to play a role in gonadal development although its function is controversial. It may also be involved in the control of circadian rhythms of the body.

C. Thyroid gland - responsible for release of thyroid hormone, which influences the rate of body metabolism.

D. Parathyroid glands - four glands that lie posterior to the thyroid gland and are important in the control of calcium metabolism. Sometimes, they may lie abnormally in the superior mediastinum, causing difficulty for the surgeon searching for a parathyroid tumor.

E. Suprarenal (adrenal) gland - has numerous functions. Its medulla produces adrenalin (epinephrine) and noradrenalin (norepinephrine) which, among other things, activate the sympathetic system. The adrenal cortex produces hormones (e.g. aldosterone) that influence the balance of electrolytes and water in the body; hormones (e.g. cortisol) that affect the immune response and the metabolism of carbohydrates; and estrogenic and androgenic hormones.

F. Pancreas - apart from its digestive function, the pancreas produces hormones (glucagon and insulin) that help regulate the metabolism of carbohydrates.

G. Ovary - produces ova and female sex hormones.

H. Testes - produces spermatozoa and male sex hormones.

I. Thymus. Its hormonal function is questionable and it may more properly be considered a part of the lymphatic system as it functions in lymphocyte production and the immune response.

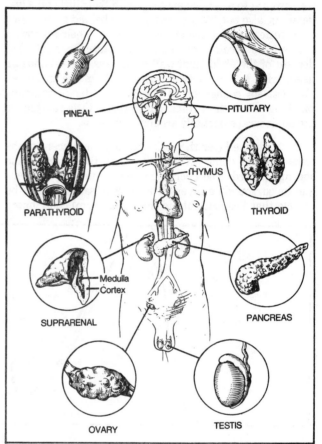

Figure 10-1

CHAPTER 11. THE GENITOURINARY SYSTEM

Fig. 11-1. Fetal migration of the testes. Once upon a time, in the embryo, the testes were lonely, lying separated from one another on either side of the abdomen just outside the future peritoneum. They began to migrate to the scrotum to meet one another, with their little tails (vas deferens, also outside the peritoneum) behind them. "Stop those testes!", called the abdominal wall. "Their migration will set up weak spots that may cause hernias!" The deepest of the abdominal wall structures, the peritoneum itself, tried to evaginate and grab hold of the testes but alas lost its connection to the rest of the peritoneum, pinching off as the **tunica vaginalis**, a little sac that in the adult wraps partially around each testis.

Each muscle layer then tried to stop the descent of the testes and, in doing so, contributed to the layers of the spermatic cord. The spermatic cord is the vas deferens and its surrounding layers. The transversalis muscle (the muscle that lies just outside the peritoneum) contributed its fascia, which clung to the vas deferens (forming the **internal spermatic fascial layer** of the spermatic cord) but this did not stop the descent. How much can loose fascia do anyway? The internal oblique muscle contributed cremasteric muscle fibers, to form the **cremasteric layer** of the spermatic cord. The cremasteric muscle pulled but could only cause partial withdrawal of the testes (the cremaster reflex, fig. 4-43). The external oblique contributed the **external spermatic fascia** (fig. 11-2).

Despite all this effort, the testes remained descended, leaving a large gap in the muscle layers of the abdominal wall. The gap was especially large in the transversus muscle, but was largely closed by the internal and external obliques. These muscles surrounded the spermatic cord, forming an inguinal canal figs. 11-3 and 11-4).

Fig. 11-2. Layers of the spermatic cord and scrotal wall as seen in cross section, above the level of the testis.

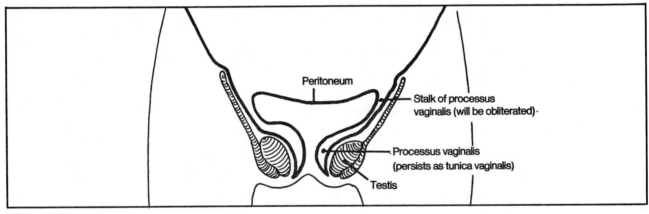

Figure 11-1

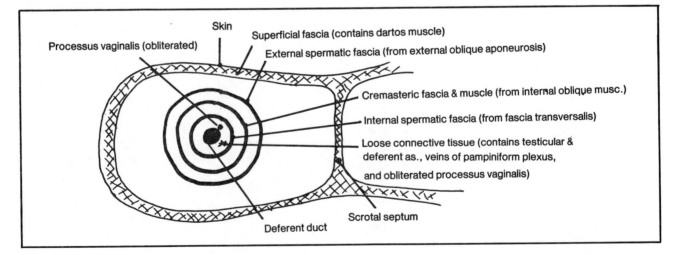

Figure 11-2

Fig. 11-3. Muscles forming the inguinal canal. The **transversalis fascia** overlies the peritoneum. The internal (deep) inguinal ring is simply an opening in the transversalis fascia through which the spermatic cord extends. The external (superficial) inguinal ring is formed by the border of the external oblique muscle.

Fig. 11-4. Mnemonic for the inguinal canal. Imagine the spermatic cord as a necktie. The "TIE" muscles (T-Transversus abdominis; I-Internal oblique; E-External oblique) form the successive arcades that surround the spermatic cord and·constitute the inguinal canal.

The inguinal canal provides protection against inguinal herniation. Anterior to the cord, protection is provided mainly by the external oblique muscle. Posterior to the cord, protection is largely provided by the **conjoint tendon** (falx inguinalis) which is a fusion of the tendons of the internal oblique and transversus muscle (fig. 11-3). Not shown in figure 11-3 is additional reinforcement that exists posterior to the spermatic cord near the superficial inguinal ring:

a. the **reflex inguinal ligament** - an extension of the external oblique across the midline
b. the **lacunar ligament** - an extension of the inguinal ligament.

The **superficial inguinal** ring is the region (medially) where the spermatic cord emerges from all the muscles.

The **deep inguinal ring** is the deepest point (laterally) where the spermatic cord plunges deep to the transversalis fascia. This is at the point where the peritoneum had invaginated as the processus vaginalis. The inguinal canal is the passageway between deep and superficial rings.

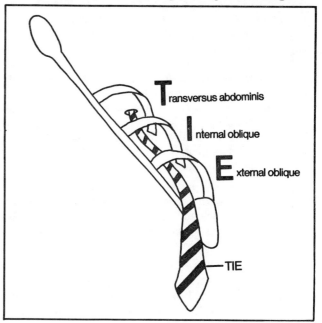

Figure 11-4

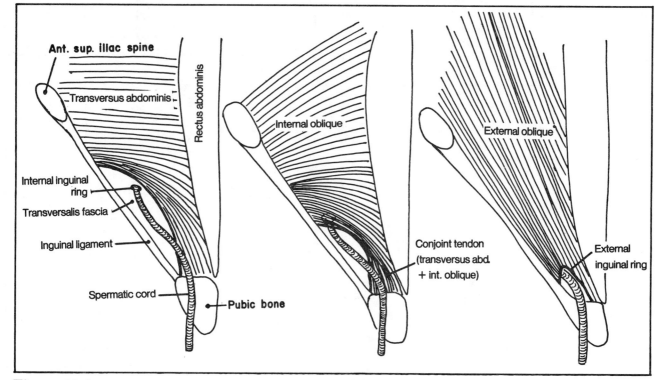

Figure 11-3

Fig. 11-5. Direct and indirect inguinal hernias. "**Indirect**" inguinal hernias take the long (indirect) route, passing through the length of the inguinal canal to the superficial inguinal ring, to the scrotum. "**Direct**" inguinal hernias take the short (direct) route by directly protruding through the region of the superficial ring, pushing against (or around) the conjoint tendon. The inferior epigastric artery lies just medial to the deep inguinal ring: i.e., between the zones of direct and indirect inguinal hernias. Palpating its pulse may be useful in telling the surgeon where the deep inguinal ring lies. **Hesselbach's triangle** is a surgically defined zone where direct inguinal hernias occur. It is bound by the inguinal ligament, inferior epigastric artery and rectus abdominis muscle. Direct and indirect inguinal hernias occur above the inguinal ligament. **Femoral hernias** occur in the femoral canal below the inguinal ligament (fig. 17-8).

In a **vasectomy** procedure, the ductus deferens is ligated. It is necessary to isolate the ductus deferens and not ligate the spermatic cord as a whole. This is because the major arterial supply to the testes lies deep within the cord and may be compromised.

Persistence of the processus vaginalis may lead to fluid collection in the scrotum (**hydrocele**) or to indirect inguinal hernia. Indirect inguinal hernias are more common than direct hernias, accounting for about 75% of inguinal hernias.

Sometimes a testicle does not descend during development. This is a surgical problem as there is an increased incidence of carcinoma developing in an undescended testicle if the condition is not corrected. Undescended testes are also frequently associated with indirect inguinal hernia.

The inguinal region in the female also contains an inguinal canal. This canal houses a **round ligament** rather than a spermatic cord. The round ligament attaches at one end to the uterus and on the other end to the labium majus (the homologue of the scrotum).

Fig. 11-6. Genitourinary system in the male (see also figs. 11-7 and 11-8).

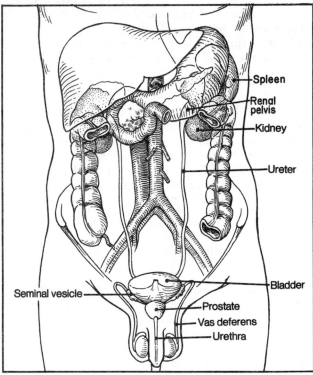

Figure 11-6

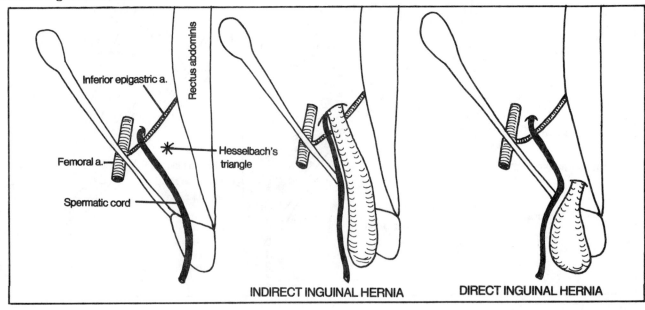

Figure 11-5

The kidney is a retroperitoneal structure. Note that the right kidney lies somewhat lower than the left kidney (the liver is in the way). Normally, only the right kidney is palpable. The lower poles of both kidneys extend below the 12th rib margin. In contrast, the spleen normally remains within the confines of the rib cage.

Ureteral stones tend to lodge in one of three places: the junction between ureter and renal pelvis; the junction between ureter and bladder; the point where the ureter crosses the external iliac artery. The ureters descend anterior to the ends of the transverse vertebral processes (L2-L5), a point useful in searching for ureteral stones on plain x-ray.

Sometimes an extra (ectopic) renal artery may compress the superior end of the ureter, causing obstruction of urinary flow. As renal arteries are end arteries (poor anastomoses), the surgeon will first compress the anomalous artery before tying it off, to determine (by observing blanching of the kidney) whether such ligation will excessively impair circulation to the kidney. If so, an artificial arterial bypass graft may be attempted.

Fig. 11-7. Genitourinary system in the male, Details of penile structure are reviewed in figures 4-37 through 4-40. If the bladder resembles a sharp-nosed leprechaun, a hose (ductus deferens, or vas deferens) is draped over his shoulder (ureter). Seminal vesicles are secretory

glands that contribute to the ejaculate and may function in the activation of sperm cells. The seminal vesicle and ductus deferens join to form the **ejaculatory duct**, which penetrates the prostate and joins the (prostatic) urethra. The prostate produces a component of the ejaculate which is secreted along many ductules into the prostatic urethra.

Fig. 11-8. Genitourinary system in the male, sagittal view. Note the peritoneum draping over the bladder like a skull cap.

The prostate has several lobes (fig 11-8):

(1) an anterior lobe - lies anterior to the urethra. Pathology is uncommon here
(2) a middle (median) lobe - lies between the urethra and ejaculatory duct. This lobe commonly enlarges in older men (**benign prostatic hypertrophy**) and may cause urinary obstruction
(3) a posterior lobe - lies posterior to the ejaculatory duct and is the area most easily felt on rectal exam. This lobe is commonly involved in prostatic carcinoma
(4) 2 lateral lobes (not shown in fig. 11-8) - form the lateral surface of the prostate. These may be considered as lateral extensions of the other lobes. Divisions between the various lobes are not clearly defined.

Various diseases of the prostate may be detected on rectal examination. The examining finger may reveal a

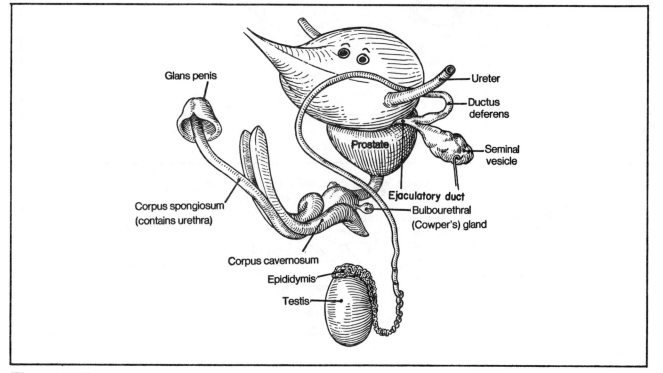

Figure 11-7

benign, enlarged prostate, a hard, nodular prostate (suggestive of malignancy), or a soft, tender prostate, as is found in **prostatitis** (inflammation of the prostate).

Fig. 11-9. The female reproductive system. Just as the scapula resembles a bull's head, the uterus resembles a ram's head, seen head on. A posterior view of the uterus is shown in order to include the ovary, which lies posterior to the broad ligament. The **broad ligament** is a fold of peritoneum that drapes over the uterus like a cape over the ram's head. It helps to support the structures that it encloses. The **suspensory ligament** is a part of the broad ligament that connects the ovary with the lateral pelvic wall and contains the ovarian vessels.

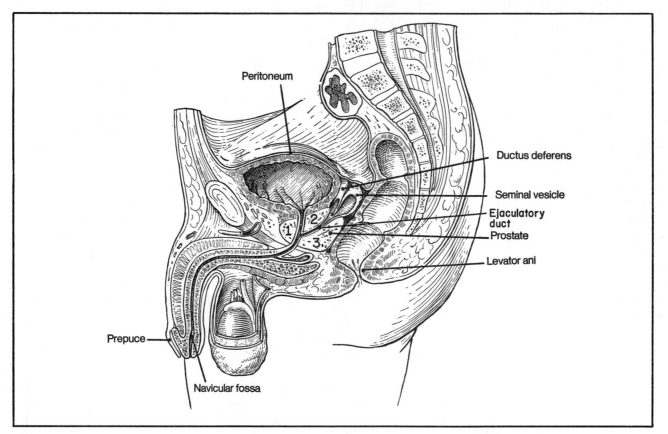

Figure 11-8

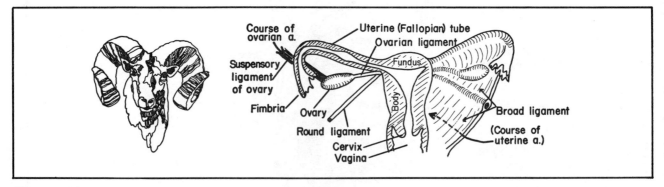

Figure 11-9

Fig. 11-10. Lateral view of the uterus, showing that the ovary lies posterior to the broad ligament, and has a mesentery (mesovarium) of its own. The ovary itself is not surrounded by peritoneum but lies directly exposed to the peritoneal cavity.

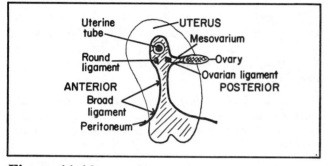

Figure 11-10

Fig. 11-11. Genitourinary system in the female, sagittal view. Note the peritoneal reflections and the **rectouterine pouch (of Douglas)** which is the lowest point in the peritoneal cavity. A needle passed through the **fornix** of the vagina into the rectouterine pouch may be used to drain fluid (e.g. blood or pus) from the peritoneal cavity. In a **culdoscopy**, a fine viewing tube is passed through this area into the peritoneal cavity for examination of the ovary and other structures. The posterior fornix may sometimes appear lumpy on pelvic examination. This most likely is stool felt through the adjacent rectal wall and should not be confused with tumor.

The relatively short urethra in the female accounts for the much higher incidence of bladder infections in women.

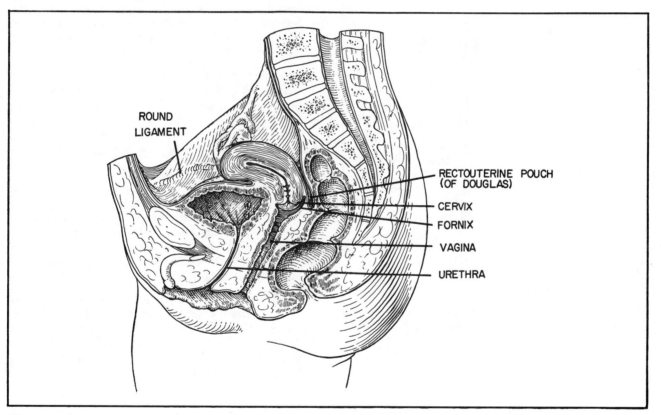

Figure 11-11

CHAPTER 12. THE NERVOUS SYSTEM

The central nervous system (CNS) includes the cerebrum, cerebellum, brain stem and spinal cord (fig. 12-1) plus a few scary-sounding structures that lie between the brain stem and cerebrum, namely the **diencephalon** and the **basal ganglia**. A detailed description of connections within the central nervous is beyond the scope of this text (fortunately).

Fig. 12-1. The central nervous system. Within the brain stem and spinal cord, the superior-inferior axis is synonymous with the rostral-caudal axis.

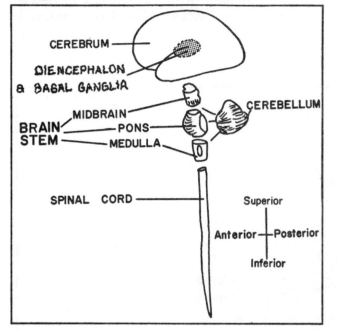

Figure 12-1

Fig. 12-2. The cerebrum.

Figure 12-2 illustrates the subdivisions of the cerebrum into frontal, parietal, occipital and temporal lobes. These

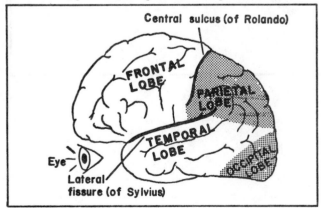

Figure 12-2

are further subdivided into bulges, called **gyri,** and indentations called **sulci** and **fissures** (small and large, respectively).

The cerebrum is shaped like a boxing glove, its anterior portion fitting into the anterior cranial fossa, the temporal lobe fitting into the middle cranial fossa, and the cerebellum fitting neatly into the posterior cranial fossa (fig. 6-41).

Fig. 12-3. Sagittal view of the brain. C.C., corpus callosum; 3V, third ventricle; P, pituitary gland; 4V, fourth ventricle. Shaded areas are zones containing cerebrospinal fluid.

The brain stem contains three parts - the **midbrain, pons** and **medulla** (fig. 12-1). The pons lies squashed against the **clivus** (fig. 12-3), a region of bone resembling a slide that extends to the **foramen magnum,** the hole at the base of the skull where the spinal cord becomes the brain stem (fig. 12-3).

Sometimes the brain stem does "slide down" the clivus, herniating into the foramen magnum. This is a serious clinical condition, generally resulting from a pressure differential between cranial and spinal cavities. Many clinicians therefore are wary in removing cerebrospinal fluid during a spinal tap in patients with high intracranial pressure.

Note the close proximity of the clivus to the nasal passages. Sometimes rare invasive tumors of the nasal passages erode and break through the clivus and damage the brain stem. Pituitary tumors may be reached surgically via the nasal passages by producing a hole in the sphenoidal bone, which houses the pituitary gland: the **"transsphenoidal approach"**.

Fig. 12-4. The cerebrospinal fluid (CSF) circulation. Arrows indicate the direction of flow of CSF. RLV, right lateral ventricle; LLV, left lateral ventricle; 3V, third ventricle; IF, interventricular foramina; AS, aqueduct of Sylvius; 4V, fourth ventricle; FL, foramen of Luschka; FMg, foramen of Magendie; CC, central canal of spinal cord.

The function of the CSF is uncertain. It may have a cushioning or a nutritive function. Within the CNS, cerebrospinal fluid is produced in chambers called **ventricles**. The walls of each ventricle contain a specialized structure called the **choroid plexus** which secretes the clear, colorless CSF. CSF flows from the two **lateral ventricles** through the two **interventricular foramina** (holes), first into the single midline **third ventricle**, from there through the single midline **aqueduct of Sylvius**, next into the single midline **fourth ventricle**, and then passes outside the brain via three openings in the fourth

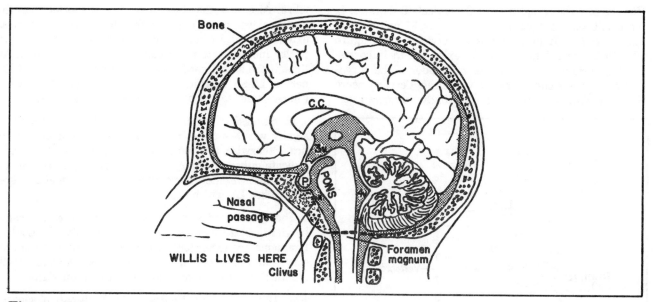

Figure 12-3

ventricle: a Middle foramen of Magendie and two Lateral foramina of Luschka. Once outside the brain stem, the CSF enters the **subarachnoid space** (the space between the arachnoid and pia) and exits into the superior sagittal sinus via specialized structures in the sinus wall called **arachnoid villi**. An obstruction at any point along this pathway will lead to dilation (expansion) of the lateral ventricles, termed **hydrocephalus**. Normally CSF does not circulate in the central spinal canal of adults, as the canal is closed off in places.

Expanded areas of subarachnoid space are called **cisterns**. The largest such cistern is the lumbar cistern, which is the site of withdrawal of CSF for examination. It is also the site of injection of various chemicals (e.g., radioopaque material for x-ray examination; spinal anesthetics; antibiotics).

The spinal cord ends at about vertebral level L2 (fig. 12-4). The subarachnoid space, as well as the arachnoid and dural membranes, end at sacral level S2. Below this point (epidural space) is the site for saddle block (epidural) anesthesia. The advantage of saddle block anesthesia is that there is no danger of it entering the CSF, spreading up the spinal canal, and paralyzing nerves along the way.

In performing a spinal tap, the iliac crest is a good landmark that approximates vertebral level L4. The tap is generally performed just above or below this level. The needle is passed between the vertebral spines and passes through the supraspinous ligament, interspinous ligament, yellow ligament, dura, and arachnoid to reach the subarachnoid space (review fig. 3-5).

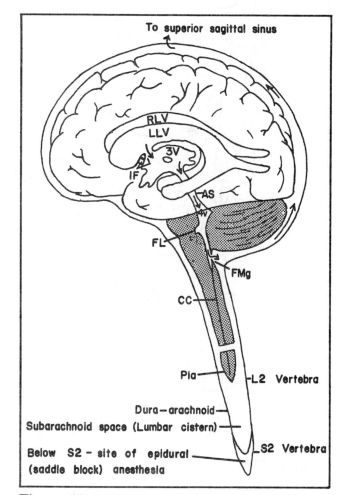

Figure 12-4

Fig. 12-5. The spinal nerves. C, cervical; T, thoracic; L, lumbar; Coc., coccygeal nerve. There are 31 pairs of spinal nerves and 12 pairs of cranial nerves. Note in figure 12-5 that cervical nerves C1-C7 exit over their corresponding vertebrae, but that the remainder of the nerves exit below their corresponding vertebrae. Cervical nerve 8 is unique, as it has no correspondingly numbered vertebra. Also, note that the spinal cord is shorter than the vertebral column. Consequently, the spinal nerve roots extend caudally when leaving the spinal cord. This disparity increases at more caudal levels of the cord. Although the spinal cord ends at about vertebral level L2, nerves L2-S5 continue caudally as the **cauda equina** ("horse's tail") to exit by their corresponding vertebrae.

In discussing spinal cord "levels", it is important to signify whether one is referring to a nerve or to a vertebra. Otherwise, a lesion of the spinal cord at "L1" might indicate either an injury of the spinal cord in the region where root L1 enters the cord, or a cord lesion at the level of vertebra L1, which lies lower.

The spinal cord is suspended within the spinal canal by **denticulate ligaments** (not shown). The denticulate ligaments are extensions of pia that lie between the sensory and motor spinal nerve roots like serrated teeth (denticulate = tooth-like), and connect with the dura.

Fig. 12-6. The spinal nerve roots. Each pair of sensory and motor nerve **roots** (**cervical, thoracic** and **lumbar**) joins in the intervertebral foramen (small of the snowman's back in fig. 2-8) to form a **spinal nerve**. Once outside the vertebral canal the spinal nerve divides into a posterior (dorsal) and an anterior (ventral) **ramus**, EACH of which contains a mixture of motor and sensory fibers. The **sacral** nerve roots, however, are so long, being part of the cauda equina, that they can't wait to get to the sacral foramina; the roots fuse and form spinal nerves while in the vertebral canal and also divide into posterior and anterior rami within the canal. Thus each sacral nerve has two exit points (not just one intervertebral foramen) - a posterior and an anterior foramen on either side of the sacrum, for the posterior and anterior rami.

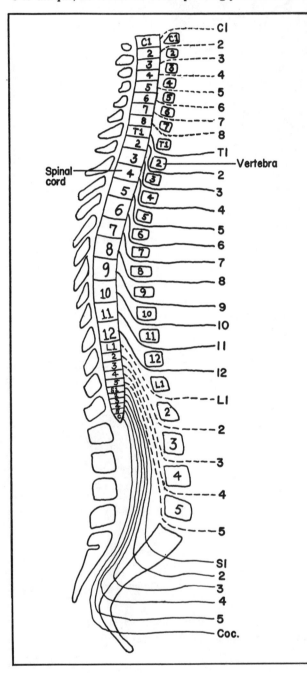

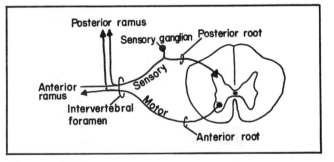

Figure 12-6

Posterior rami supply skin near the middle of the back, as well as midline skeletal muscles of the spine (fig. 12-7). **Anterior rami** supply the rest of the neck, trunk, and extremities. The cervical, brachial and lumbosacral plexuses consist of extensions of **anterior** rami.

Figure 12-5

Fig. 12-7. Distribution of the posterior rami of the spinal nerves. The darkly shaded area indicates the muscles innervated by posterior rami. These are the spinal muscles that lie near the midline (splenius, erector spinae, transversospinalis). The lightly shaded area indicates the area of cutaneous innervation by posterior rami. This is a broader area than the muscular zone.

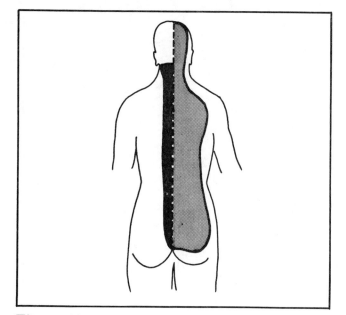

Figure 12-7

The Dermatome Map

Both posterior and anterior rami together combine to produce the dermatome map of the body (fig. 12-8). This is a map of the sensory distribution of spinal nerves C2-S5 on the skin (C1 has no sensory distribution). Generally, the motor distribution of these nerves underlies and corresponds to the skin map, with significant exceptions. For instance, the diaphragm is innervated by nerves C3-C5, which, for skin sensation, are represented on the neck and shoulder in the dermatome map.

Fig. 12-8. The dermatome map. Roughly speaking, the shape and location of key sensory dermatomes are as follows:

C1 - no sensory dermatome
C2 - a skull cap
C3 - a high collar, as found on a turtleneck sweater
C4 - a cape around the shoulder
C5-T1 - upper extremities ("thumb-suckers suck C6")
T5 - nipple
T10 - belly button
L1 - Inguinal Ligament (i.e. L1 = IL)
L4 - knee (also the nerve mediating the knee jerk reflex). L4 also provides sensation to the big toe, which kicks the examiner testing the knee jerk reflex.
L5 - three middle toes (between L4 and S1)
S1 - ankle jerk reflex (poor L5 has no reflex)
S1-S2 - predominantly on the rear of the lower extremities.
S2-S5 - genital and anal zones

The Cervical Plexus

Plexuses are networks of nerves fused together in complex ways. Excluding the autonomic nerve plexuses, the three main nerve plexuses of the body are the **cervical, brachial** and **lumbosacral plexuses**.

Fig. 12-9. The cervical plexus. The cervical plexus resembles a friendly little person whose neck and arms are the anterior rami of C1, C2, C3 and who is waving "Hi" with his two little hands. He is waving "Hi", because his legs supply the infra"HY"oid muscles. CN12, which supplies the tongue, even tries to lick the little man affectionately, running itself along the side of his body but keeping its distance above the hyoid bone along with a small branch of the plexus. Additional sensory (cutaneous) branches arise from C2, 3, 4 rami to supply C2, 3, 4 dermatomes. Cutaneous branches C2 and C3 (C1 has no sensory distribution) come off the man's shoulder and arm between C2 and C3. The little man hides behind the sternocleidomastoid muscle. Thus the cutaneous nerves from the little man have to crawl out around the posterior border of the sternocleidomastoid muscle in order to reach the skin (fig. 12-10).

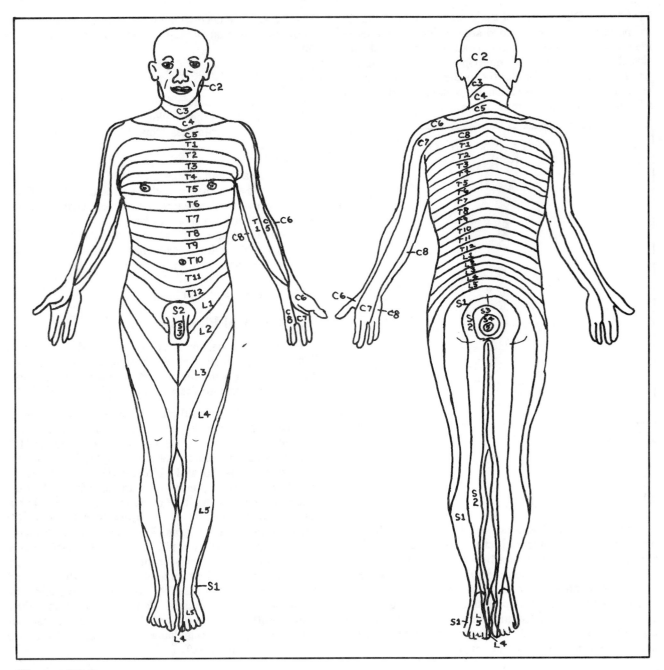

Figure 12-8

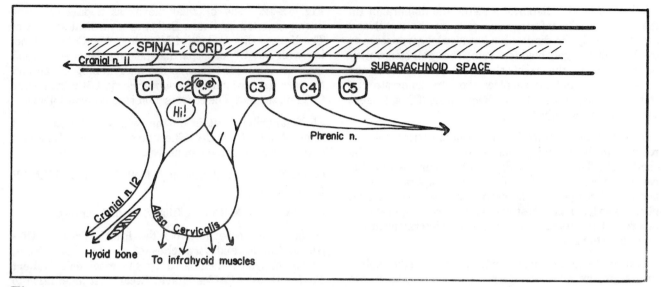

Figure 12-9

Fig. 12-10. Cutaneous branches of the cervical plexus exit around the posterior border of the sternocleidomastoid muscle, below the level of the accessory nerve (cranial nerve 11). One can anesthetize a large area of neck by performing a nerve block in this region.

Cranial nerve 11 has its origins in cervical cord segments C1-C5. Afraid of being ensnared in a complicated plexus, CN11 refuses to leave via the intervertebral foramen. Instead, it travels in the subarachnoid space up through the foramen magnum, touches the vagus nerve within the skull and then comes back down through the

jugular foramen to innervate the trapezius and sternocleidomastoid muscles (figs. 12-9 and 14-4).

The **phrenic nerve** (C3, 4, 5) also is wary of plexuses. Rather than get involved with the little man, it moves in a direction opposite to that of cranial nerve 11, toward the diaphragm (fig. 12-9).

Fig. 12-11. Course of the phrenic nerve. It lies in the neck and thorax somewhat lateral to the vagus nerve (fig. 14-14), traveling around the lateral border of scalenus anterior, then extending between the subclavian artery and vein to enter the thorax. It runs in "phront" of the root of the lungs, whereas the vagus nerve runs posterior (fig. 14-14). It follows the pericardium to the diaphragm, which it supplies from above and below. Branches ex-

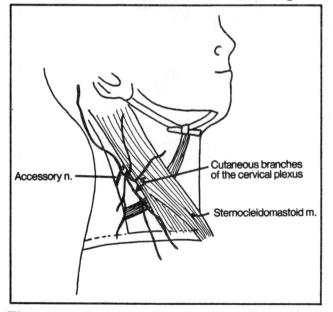

Figure 12-10

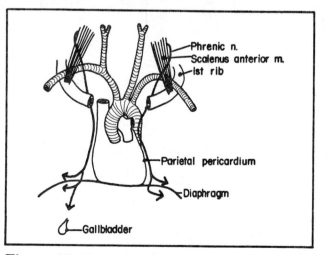

Figure 12-11

tending below the diaphragm pierce the diaphragm (in the region of the inferior vena cava on the right) and supply not only the diaphragm but nearby peritoneum to the level of the gall bladder. Pain from gall stones may be referred to the right shoulder, because both the phrenic nerve and the shoulder area are innervated by spinal cord segments C3, 4, 5. (Remember, "C3, 4, 5 keeps the phrenic nerve alive!").

The pain of pleurisy (pleural inflammation) may be referred to the overlying shoulder or to local areas of the chest wall. These two kinds of pain occur because the parietal pleura overlying the medial diaphragm is innervated by the phrenic nerve, whereas the parietal pleura overlying the chest wall and lateral diaphragm is innervated by intercostal nerves (The visceral pleura is insensitive to pain).

Referred pain phenomena may occur with other spinal cord segments. Cardiac referred pain may occur in the left upper extremity, as spinal segment T1 supplies both the heart (via sympathetic nerves) and the upper extremity. Pain from the appendix (T10) may be referred to the umbilicus (dermatome T10). Colic of the ureter (T11 to L2) may radiate to the groin along the inguinal ligament (L1).

Fig. 12-12. The brachial plexus as a dancing scene (idea suggested by Susan Balke, as a medical student at The Kansas City College of Medicine). Mr. C5C6 is dancing with Ms. C7, each partner holding the other around the rear. Mr. C8T1 tries to cut in, grabbing C7 around the rear (i.e., around the posterior cord). C5C6 then kicks C8T1 on the foot. The conversation is roughly as follows:

C5C6 (angrily): "Cut in on me, will you! I'll show you some MUSCLE (MUSCULocutaneous nerve)! Take that ("Kick" - region of median nerve)!."

C8T1: "Yow! You broke my foot! YOU'LL (ULNAR nerve) be sorry."

C7: "Go get an X-RAY (AXillary, RAdial nerves).

The above scene describes the **Rami, Trunks, Divisions, Cords,** and **Branches** of the brachial plexus, which may be remembered by the mnemonic **Robert Taylor Drinks Coffee Black.** Recall that in addition to the above segments, which lie outside the intervertebral foramina, there are also sensory and motor **Roots,** which lie within the spinal canal. The roots fuse at about the level of the intervertebral foramina, forming the spinal nerve which divides into anterior and posterior rami (fig. 12-6). The brachial plexus originates from anterior rami.

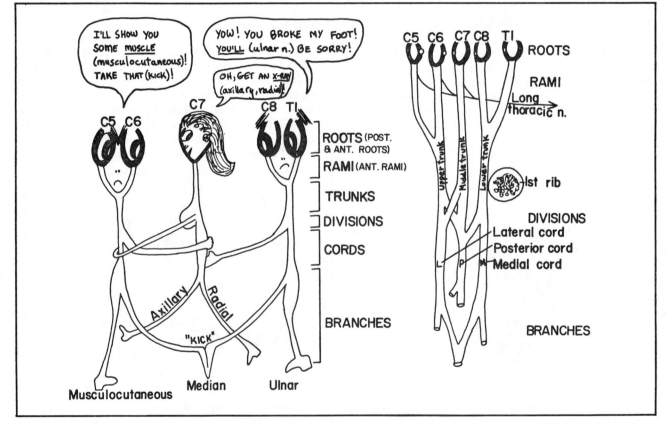

Figure 12-12

Some references call the rami "roots", but this usage is confusing. In the above figure, the pupils of the eyes are posterior (dorsal) root ganglia.

In figure 12-12, the first rib is seen in cross section, and looks like a cross section through a tree "trunk". The (lower) "trunk" of the brachial plexus overlies the first rib.

Fig. 12-13.   Brachial plexus lesions.

Lesion #1 (in fig. 12-13). **Waiter's tip injury.** A waiter is tapping (injuring) Mr. C5C6 on the shoulder at point #1, asking for a tip. This results in the "waiter's tip" injury. The patient's hand (note the right hand of the waiter) is held as if asking for a tip. This posture occurs for at least two reasons: The shoulder becomes inturned as the infraspinatus muscle is denervated. The nerve to the latter muscle (suprascapular nerve) comes off the upper trunk. As the infraspinatus muscle rotates the humerus laterally, its non-functioning leads to inturning of the shoulder. Secondly, the biceps muscle, innervated by the musculocutaneous nerve, serves partly as a supinator of the forearm (fig. 4-6A). Its nonfunctioning leads to pronation. Hence, shoulder inturning + pronation = waiter's tip posture. There will also be loss of skin sensation in a C5C6 distribution. Lesion #1 may occur with stab wounds of the neck; in birth injuries, or in falls (as from a motorcycle) where the angle between the shoulder and neck is suddenly widened. Loss of sensation or

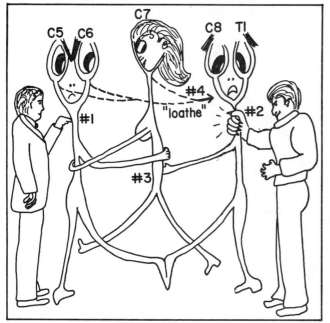

Figure 12-13

weakness near the middle of the back (dorsal ramus distribution), would suggest that the roots of C5 and/or C6 have been pulled out, thereby severing the posterior rami as well.

Lesion #2 (in fig. 12-13). **Clawhand.** A giant claw (belonging to the dance hall bouncer) is attempting to remove Mr. C8T1 from the ballroom by grabbing him around the neck (nerve trunk) at point #2. This injury results in a "claw hand" (see fig. 12-17). This occurs in part because the nerve fibers going through this trunk supply all the small muscles of the hand, most particularly the lumbricales. In figure 4-13 review the "L"-shaped position of the hand when the lumbricales are contracting and note that the claw hand (fig. 12-17) is in a sense the opposite of the "L", as the lumbricales are not functioning. There will also be impairment of wrist flexion (the wrist becomes extended) as the wrist flexors are weak. (The wrist extensors are supplied by the radial nerve, which remains unaffected).

Injury #2 may occur with compression by an (extra) cervical (C7) rib; by sudden pulling of the arm as in trying to grab onto an object to break one's fall; as a birth injury; by compression from lymph node metastases. Sensory defects will occur in a C8T1 distribution.

Lesion #3 (in fig. 12-13). **Wrist drop.** As Mr. C8T1 reaches for C7's rear(point #3), dangerously close to her CROTCH (a CRUTCH applying too much pressure under the arm might produce such an injury), C7's wrist will soon drop to brush off C8T1's hand. Such an injury produces a wrist drop because of loss of the extensors of the wrist, which are supplied by the radial nerve. The patient has difficulty in extending the wrist or fingers, or the elbow. In addition, there may be some sensory loss (see fig. 12-16) but this often is slight or absent as adjacent nerves overlap the radial nerve in skin innervation.

Lesion #4 (in fig. 12-13). **Scapular winging.** Note one additional feature of this dancing scene. Mr. C5C6 and his girl friend, C7, LOATH (LOng THoracic nerve) Mr. C8T1 and are looking daggers at him. This is the long thoracic nerve, or nerve to serratus anterior(C5-C7), which innervates the serratus anterior muscle. Damage to this nerve results in "scapular winging"; i.e., the scapula protrudes when the patient pushes against a wall (fig. 4-3) because the serratus anterior muscle normally holds the scapula against the rib cage. If scapular winging is present, this suggests that the plexus injury may be relatively close to the vertebral column. Alternatively, the long thoracic nerve may be injured along its course over the thorax, as might occur during a radical mastectomy.

Fig. 12-14. Relations of the brachial plexus and first rib.

1. The brachial plexus passes between the first rib and clavicle.
2. The 3 trunks of the plexus lie at about the level of the first rib (fig. 12-14).
3. The plexus rami lie in the neck, whereas the divisions and cords lie in the axilla (armpit). The 5 main nerve branches of the upper extremity begin at the upper arm.

The brachial plexus, subclavian artery and subclavian vein all travel between the clavicle and first rib. However, only the brachial plexus and subclavian artery travel between the anterior and middle scalene muscles. The subclavian vein travels superficial to the anterior scalene muscle. It is fortunate that the subclavian artery and plexus lie away from the vein, for in catheterizing the subclavian vein, one does not wish to puncture the artery or plexus. One must be careful, however, not to puncture the lung, which underlies the subclavian vein. Needle puncture of the subclavian vein is performed under the clavicle.

The lower brachial plexus trunk and the subclavian artery circulation may be compromised if there is excessive hypertrophy of the anterior scalene muscle (**scalenus anticus syndrome**)(see fig. 12-14). Compromise of the lower trunk and subclavian artery may also occur if the latter drape over an extra (cervical) rib. These problems often can be corrected surgically by severing the scalenus anterior muscle at its insertion or removing the extra rib.

The musculocutaneous, median, ulnar, axillary and radial nerve branches each contain mixtures of nerve fibers arising from several nerve roots. Injuries to these nerves, therefore, will give rise to different sorts of clinical problems than do injuries to nerve roots, or regions of the brachial plexus. The sensory skin distribution of individual nerves overlap one another and nerve lesions may result in patchy areas of sensory loss rather than in a neat dermatome distribution.

Fig. 12-15. Nerves of the upper extremity.

A. The **axillary nerve** lies at the level of the axilla. It may be injured with shoulder dislocations, or fractures of the surgical neck of the humerus. There is paralysis of the deltoid with difficulty elevating the arm, as well as anesthesia of the lateral shoulder.

The **radial nerve** commonly is injured by compression, as in "Saturday night palsy". The patient typically falls asleep in a chair, drunk, after an active Saturday night out, with the arm over the back of a chair, thereby compressing the radial nerve. The result is a wrist drop. Alternatively, fractures of the humerus may injure the radial nerve since this nerve lies in the radial sulcus, which indents the humerus. Fracture below the midshaft of the humerus spares the ability to extend the elbow, as

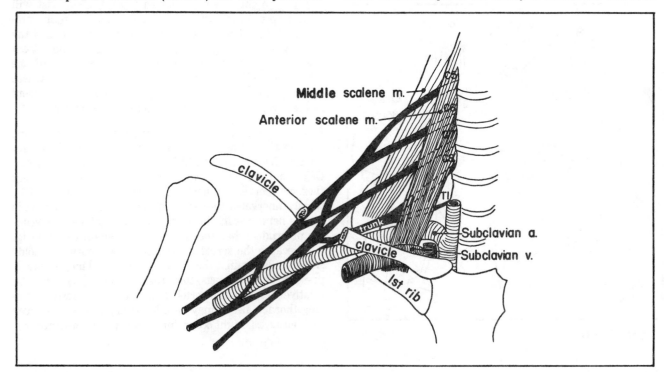

Figure 12-14

the nerve to the triceps comes off the radial nerve high up on the humerus (fig. 12-15). Note that the radial nerve does not innervate any intrinsic hand muscles.

B. The **musculocutaneous nerve** may be injured by a laceration, leading to difficulty in flexing the elbow (biceps) and variable sensory loss.

C. The **median nerve** assists in flexion of the wrist and fingers. The finger flexion that it controls occurs mainly through the forearm muscles, rather than the intrinsic hand muscles. It does, however, have an important motor function within the hand where it becomes the $1,000,002 nerve ($2, because of its relatively minor function of innervating 2 lumbrical muscles; $1,000,000, because it innervates the vital thenar muscles, which are critical in the functioning of the thumb. Sever this nerve and you may be liable to a $1,000,000 lawsuit). Deficits following lesions of the median nerve will vary according to the lesion site. Above the elbow there will be significant weakness of wrist and finger flexion. With laceration at the wrist, as in a suicide attempt, or in **carpal tunnel syndrome**, there will be little deficit in wrist flexion; the

major motor loss will be that of movements of the thumb.

D. The **ulnar nerve** is mainly concerned with the function of the intrinsic muscles of the hand. Fifteen of the 20 intrinsic hand muscles are controlled by the ulnar nerve. The other 5 are controlled by the median nerve. Ulnar nerve lesions may occur with fractures of the elbow. The ulnar nerve is the nerve involved when one hits the "funny bone", i.e., strikes the elbow on a hard object. There is an immediate tingling sensation over the medial 1 1/2 fingers, the sensory distribution of the nerve. In more severe injuries, a claw hand may result. The "claw" is not as pronounced as the claw of the ballroom bouncer in lower brachial plexus trunk lesions as lumbricals 2 and 3 (innervated by the median nerve) are spared in ulnar nerve injuries (fig. 12-17). With ulnar nerve injuries at or above the elbow, there will also be difficulty with wrist flexion; in addition, the wrist will tend to drift radially on attempted flexion as the flexor carpi ulnaris is malfunctioning.

The intrinsic hand muscles are mainly innervated through spinal cord segment T1.

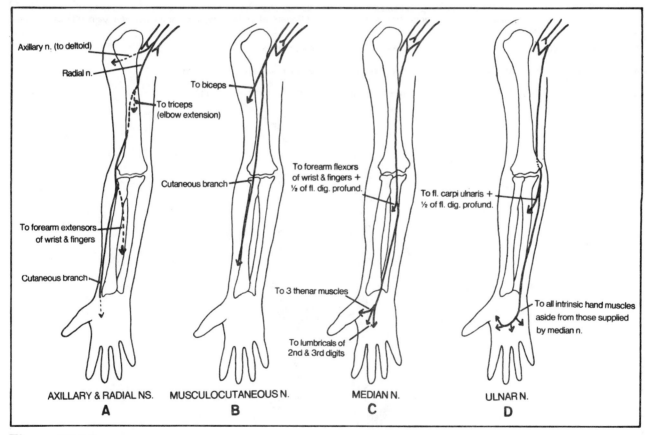

Figure 12-15

Fig. 12-16. Motor and sensory deficits following nerve injuries in the upper extremity.

Fig. 12-17. The ulnar claw hand in ulnar nerve lesions compared with the claw in brachial plexus (or combined ulnar and median nerve) injuries. Although the claw is not as pronounced in ulnar nerve lesions, this, nonetheless, is a severe deficit. There is pronounced loss of the ability to adduct the fingers; the patient cannot grasp tightly a piece of paper placed between the fingers, because all the interosseus muscles are paralyzed. The ulnar nerve also supplies sensory fibers to 1 1/2 fingers as well as motor fibers to 1 1/2 muscles in the forearm (figs. 12-15 and 12-16).

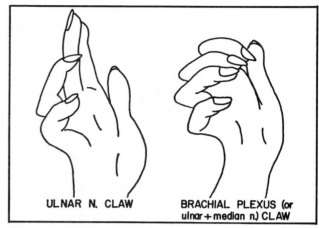

ULNAR N. CLAW   BRACHIAL PLEXUS (or ulnar + median n.) CLAW

Figure 12-17

NERVE	MOTOR FUNCTIONS IMPAIRED WITH INJURY
RADIAL (C5-C8)	Elbow and wrist extension (patient has wrist drop); extension of fingers at metacarpo-phalangeal joints; triceps reflex.
MEDIAN (C6-T1)	Wrist, thumb, index,, and middle finger flexion; thumb opposition, forearm pronation; ability of wrist to bend toward the radial (thumb) side; atrophy of thenar eminence (ball of thumb).
ULNAR (C8-T1)	Flexion of wrist, ring and small finger (claw hand); opposition little finger; ability of wrist to bend toward ulnar (small finger) side; adduction and abduction of fingers; atrophy of hypothenar eminence in palm (at base of ring and small fingers).
MUSCULOCUTANEOUS (C5-C6)	Elbow flexion (biceps); forearm supination; biceps reflex.
AXILLARY (C5-C6)	Ability to move upper arm outward, forward, or backward (deltoid atrophy).
LONG THORACIC (C5-C7)	Ability to elevate arm above horizontal (patient has winging of scapula).

AXILLARY

RADIAL

MUSCULO-CUTANEOUS

RADIAL

ULNAR   MEDIAN   ULNAR

Figure 12-16

Injuries To the Cervical Cord

Traumatic injuries to the cervical cord result in profound neurologic deficits. The functional outcome differs vastly depending on the precise level of injury. Apart from sensory disturbances, which follow a dermatome distribution, the following functional motor losses may occur:

C3 LEVEL (cervical cord severed, leaving roots of C3 and above intact): total quadriplegia (cannot move any extremity); requires respirator, as diaphragm is largely innervated by C4.

C4 LEVEL (roots of C4 and above intact): patient can shrug shoulders only; respirations weak but patient can live without respirator.

C5 LEVEL (roots of C5 and above intact): patient can move shoulder (deltoid - C5, 6) and flex elbow (biceps - C5, 6), but these actions are weak (C6 is lost). The patient cannot move a wheelchair independently.

C6 LEVEL (roots of C6 and above intact): The patient can move a wheelchair independently, but cannot grasp objects.

C7 LEVEL (roots of C7 and above intact): The patient can barely hold objects.

C8 LEVEL (roots of C8 and above intact): Long finger flexors are now functioning but grasp is still impaired because of paralysis of intrinsic hand muscles. The hand has a clawed appearance (lumbricales malfunctioning).

T1 LEVEL (roots of T1 and above intact): Upper extremity functioning intact, but patient is paraplegic (cannot move lower extremities).

Intercostal Nerves

Fig. 12-18. The intercostal nerves. Continuing along the spinal column, below the level of the brachial plexus, are spinal nerve roots T1-L2, which supply the trunk. These nerves are relatively simple and have straightforward courses with no plexuses. As do other nerves, they form posterior and anterior rami which contribute to the dermatome pattern. The anterior rami form the intercostal nerves. Intercostal nerve T1 (and sometimes T2) does supply part of the brachial plexus. Otherwise, T1-T12 travel in the intercostal spaces in the same plane as the intercostal arteries - between the internal and innermost intercostal muscles. As for the arteries, the nerves do not like to travel between the external and internal intercostal muscles, because these muscles run at oblique angles to one another and this is confusing to the nerves.

T12, the last intercostal nerve, cannot lie between two ribs as it lies under the 12th rib. Thus, it is called the subcostal nerve, and it supplies part of the abdominal wall (fig. 12-19). In fact, nerves T7-T11, after running out of rib cage, also supply part of the abdominal wall. These nerves, as well as the lower abdominal nerves, run between the internal oblique and tranversus abdominis muscles which are analogous with the internal and innermost intercostal muscles.

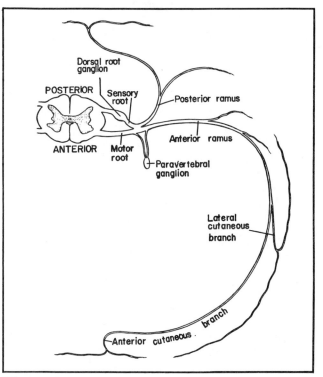

Figure 12-18

Fig. 12-19. Distribution of the subcostal(T12), iliohypogastric(L1), ilioinguinal(L1), genitofemoral(L1,L2) and lateral femoral cutaneous(L2,L3) nerves. These originate behind the psoas major muscle. The lower abdominal wall (skin and muscle) is supplied by spinal nerve L1. Recall that the inguinal ligament lies at about dermatome L1 (fig. 12-8). Nerve L1 divides into an **iliohypogastric** and an **ilioinguinal nerve**. They go where their names imply. The iliohypogastric ends in the skin overlying the pubis. The ilioinguinal covers somewhat lower territory, passing through the superficial inguinal ring to reach the skin at the base of the penis and scrotum (or base of the clitoris and labia majora).

The iliohypogastric and ilioinguinal nerves may be damaged during an appendectomy. This may predispose to a direct inguinal hernia, because of laxity of the muscles supplied by these nerves.

Pain in the inguinal region (L1) may have a local origin (such as an inguinal hernia) or may be referred from irritated L1 roots, or from a ureter (T11-L2) that is blocked by a stone.

The **genitofemoral nerve** (L1,L2 - fig. 12-19) goes to the genitals and the femoral area of the thigh. Realizing that it has no obligation to innervate the abdomen, it does not follow the route of passing between the internal oblique and transversus abdominis muscles. The best way for it to reach the anterior thigh is to follow the psoas major muscle, which passes under the inguinal ligament.

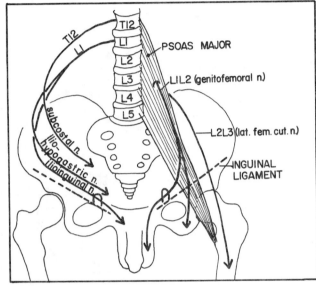

Figure 12-19

On the way, it sends a branch through the inguinal canal that reaches the scrotal skin (labium majus) and cremaster and dartos muscles.

The **lateral femoral cutaneous nerve**, derived from L2 and L3 anterior rami, extends inferiorly on the iliacus muscle (fig. 4-44). It supplies thigh skin lateral to that of the genitofemoral nerve (consider the dermatome map of L1, L2 in fig. 12-8). The lateral femoral cutaneous nerve also passes under the inguinal ligament but so far laterally that it sometimes can be pinched off by an excessively obese overhanging abdomen, thus causing **meralgia paresthetica**, an uncomfortable burning, or otherwise discomforting sensation, in the lateral femoral area, from irritation of the lateral femoral cutaneous nerve.

The lateral femoral cutaneous nerve may be injected near the anterior iliac spine to produce local anesthesia for obtaining a skin graft.

Lumbosacral Plexus

The lumbosacral plexus innervates the skin and skeletal muscles of the lower extremity and perineal area (figs. 12-20 and 12-21). Like the cervical and brachial plexuses, its nerve fibers are extensions of anterior rami.

Fig. 12-20. The lumbosacral plexus presented as a mad rescue scene. Elsie (L3) is trying to rescue her clumsy boyfriend Slim (S1) from a septic tank (SciaTIC nerve), using a rope and a balloon. What a mess! He has some

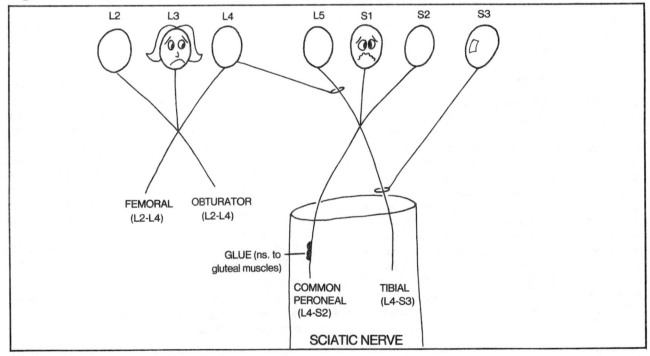

Figure 12-20

GLUe (nerves to GLUteus muscles) on his leg. She is pregnant - is FEMale (FEMoral nerve) and has an OBstetric condition (OBturator nerve)(Slim's fault).

Of the gluteal nerves, the **inferior** gluteal nerve supplies gluteus **maximus**, whereas the **superior** gluteal nerve supplies gluteus **medius and minimus**. Injury to the superior gluteal nerve (e.g., direct trauma, polio) results in the "gluteus medius limp": the abductor function of gluteus medius is lost and the pelvis tilts to the unaffected side when the unaffected extremity is lifted on attempting to walk. To some, the gait resembles a sexy wiggle.

Fig. 12-21. Escape of the lumbosacral plexus from the pelvis. The **femoral nerve** (corresponds to the femoral artery) exits the pelvis under the inguinal ligament. The **obturator nerve** (corresponds to the obturator artery) leaves through the obturator foramen. The **sciatic nerve** and the **pudendal nerve** (S2, 3, 4) leave the pelvis via the greater sciatic foramen. The sciatic nerve does not have a major corresponding artery except where it enters the popliteal fossa (behind the knee) to accompany the popliteal artery. Actually, the sciatic nerve may bleed profusely when cut in the thigh, because hidden within it is a long branch of the inferior gluteal artery (the inferior gluteal artery comes off the internal iliac artery).

The pudendal nerve corresponds to the internal pudendal artery in that after leaving the pelvis through the greater sciatic foramen the pudendal nerve returns, via the inferior sciatic foramen to run under the levator ani muscle and innervate the perineum. The pudendal nerve runs in the ischiorectal fossa within a pudendal canal that lies against the obturator internus muscle (fig.4-41). Damage to the pudendal nerve may occur during surgery, as for an ischiorectal abscess, leading to important motor and sensory losses. Branches and functions of the pudendal nerve are:

(A) Inferior rectal nerve (fig. 12-21) - goes to the anal triangle. It innervates the **external anal sphincter** (skeletal muscle), not the internal sphincter which is smooth muscle and supplied by autonomic fibers. It also supplies skin around the anus and the mucous membrane of the lower aspect of the anal canal. The itch of external hemorrhoids is conveyed along this nerve. Damage to the inferior rectal nerve may result in incompetence of the external anal sphincter.

(B) Perineal nerve - goes to the urogenital triangle. It innervates the muscles of the perineum, including part of the external anal sphincter and the levator ani. The levator ani (S3,4), while innervated by the perineal nerve, is also innervated by other branches of the sacral

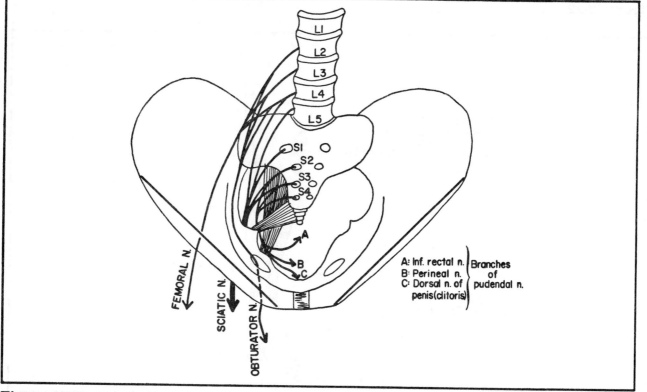

Figure 12-21

plexus (this important muscle does not take chances). The perineal nerve also supplies branches to skin overlying the muscles that it innervates.

(C) Dorsal nerve of penis (clitoris) - supplies skin of the penis (clitoris) and part of the scrotum (labia majora).

The pudendal nerve is commonly anesthetized during childbirth in order to perform an episiotomy procedure. The needle is passed through the perineal skin to a point just medial to the ischial tuberosity, where the pudendal nerve lies.

Note that all the discussion thus far has not mentioned innervation of the deeper organs of the thorax, abdomen, and pelvis. This is performed by the autonomic nervous system, which innervates smooth muscle, cardiac muscle, and glands (chapter 13).

Innervation Of the Lower Extremity

Fig. 12-22. Motor deficits of nerves of the lower extremity.

Fig. 12-23. Sensory deficits of nerves of the lower extremity.

Fig. 12-24. Innervation of the lower extremity. Innervation of the lower extremity occurs through the nerves that represent Elsie's and Slim's lower extemities, i.e., the obturator, femoral, common peroneal, and tibial nerves (fig. 12-20).

The **obturator nerve** (fig. 12-21), after passing through the obturator foramen is in a good position to innervate the (medial) adductors of the thigh. If it is injured (e.g. pelvic tumors) there is difficulty in adduct-

MOTOR FUNCTIONS IMPAIRED WITH INJURY	
FEMORAL (L2-L4)	Knee extension; hip flexion; knee jerk.
OBTURATOR (L2-L4)	Hip adduction (patient's leg swings outward when walking).
SCIATIC (L4-S3)	Knee flexion plus other functions along its branches—the tibial and common peroneal nerves.
Tibial (L4-S3)	Foot inversion; ankle plantar flexion; ankle jerk.
Common peroneal (L4-S2)	Foot eversion; ankle and toes dorsiflexion (patient has high slapping gait owing to foot drop). This nerve is very *commonly* injured.

Figure 12-22

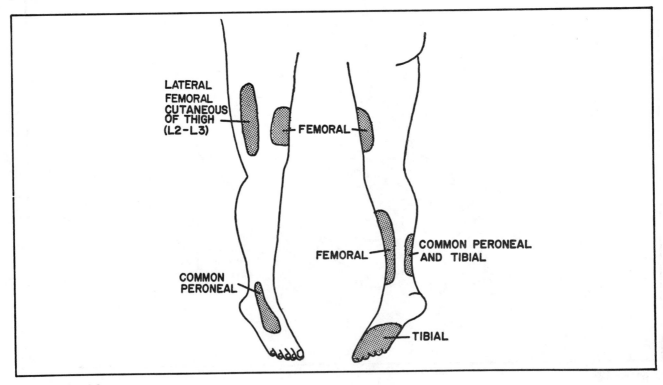

Figure 12-23

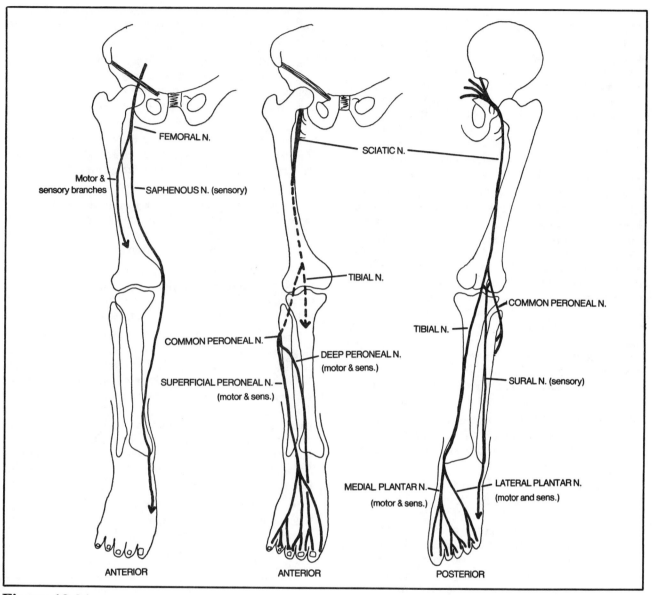

Figure 12-24

ing the thigh. The patient may retain partial adductor function as the big adductor magnus muscle is partially innervated by the sciatic nerve, and the little pectineus, also an adductor muscle, is innervated by the femoral nerve.

The **femoral nerve** (fig. 12-24) travels under the inguinal ligament, just as the iliopsoas muscle does; in fact, the femoral nerve innervates the iliopsoas muscle as well as the anterior muscles of the thigh. Thus, a femoral nerve injury will cause weakness of knee extension and, if high enough, weakness in hip flexion. There may also be decreased sensation in the anterior and medial thigh. A femoral nerve lesion may be accompanied by death

when injured in the thigh as this nerve accompanies the femoral artery, which bleeds profusely when lacerated. The femoral nerve is nervous about this dangerous arterial relationship. It breaks away from the femoral artery, leaving its less important sensory branch, the **saphenous nerve**, to cautiously accompany the femoral artery down the thigh under the sartorius muscle (fig. 17-8). However, on seeing the size of the huge, imposing sciatic nerve, the saphenous nerve absolutely refuses to follow the femoral artery through the adductor hiatus into the popliteal fossa behind the knee. (The sciatic nerve is the largest nerve in the body and occupies the popliteal fossa). The saphenous nerve, instead, switches over to accompany a less dangerous structure - the saphenous vein - and accompanies it down the medial aspect of the foot as a cutaneous nerve, ending over the big toe. Injury to the saphenous nerve may lead to numbness of skin on the medial aspect of the leg (fig. 12-24). As the saphenous nerve accompanies the saphenous vein over the medial malleolus, the nerve is in danger of being ligated during a venous cutdown in this area (see fig. 6-43). This may lead to pain along the medial aspect of the foot. If you wish to remember the distribution of the femoral nerve (L2,3,4) recall the approximate position of dermatomes L2,3,4, which approximately overly the nerve.

The **sciatic nerve** (fig. 12-24) is a combination of the **tibial (medial popliteal)** and **common peroneal (lateral popliteal)** nerves. It runs down the back of the thigh to supply the hamstring muscles. It is also responsible not only for knee flexion but for ALL the movements of the foot and toes, including foot inversion and eversion, and ankle and toe plantar flexion and dorsiflexion. These actions occur by way of the common peroneal and tibial nerves, which separate at about the level of the popliteal fossa. The cowardly femoral nerve functions simply as a cutaneous nerve (saphenous nerve branch) in the leg and foot. This **leaves the sciatic nerve with the enormous responsibility of performing for the muscles of the leg and foot what the combined median, ulnar, and radial nerves do for the forearm and hand.**

The COMMON PERONEAL nerve branch of the sciatic nerve corresponds to the RADIAL nerve. It supplies muscles and skin on the anterior (extensor) surface of the leg and foot.

The **common peroneal nerve** is a very **commonly** injured nerve. It runs down the lateral aspect of the leg and is vulnerable to tight casts or to traumatic blows from the side. The common peroneal nerve, by its lateral positioning, controls foot eversion (the tibial nerve causes inversion). Blows to the lateral leg, where the common peroneal nerve travels around the neck of the fibula, may damage the common peroneal nerve. This may cause a foot drop and difficulty with ankle eversion. Thus, the common peroneal nerve controls both foot eversion and ankle and toe dorsiflexion. The tibial nerve controls foot inversion and ankle and toe plantar flexion.

The TIBIAL NERVE remains on the posterior aspect of the leg as a straight extension of the sciatic nerve. Its MEDIAL PLANTAR division corresponds to the MEDIAN nerve, supplying muscles of the big toe and cutaneous sensation to the medial 3 1/2 toes. The LATERAL PLANTAR DIVISION of the tibial nerve corresponds to the ULNAR nerve, supplying the remaining muscles of the sole of the foot and cutaneous sensation to the lateral 1 1/2 toes.

The **sural nerve** is a cutaneous branch off the tibial and common peroneal nerves that supplies sensation to

the lateral posterior aspect of the leg and foot. It apparently is not vitally important, as surgeons sometimes use it for nerve grafts.

Fig. 12-25. Correct site for hip injection. Injection outside the upper outer quadrant may lead to injury of the sciatic nerve. The most prominent manifestation of this is a **foot drop** from damage to the peroneal nerve component of the sciatic nerve.

Usually the sciatic nerve passes inferior to the piriformis muscle. Occasionally it passes above, in which case injections into the upper outer quadrant may injure the nerve.

In view of the potential of injury to the sciatic nerve, injections in infants commonly are not given in the gluteal region but in the anterolateral thigh. Intramuscular injections also are commonly given in the deltoid muscle.

Most of the branches of the sciatic nerve, in the thigh, come off its medial side (fig. 12-24). Therefore, it is safer surgically to approach the hip joint from a point lateral to the sciatic nerve, than from its medial side.

The innervation of the toenail bed resembles that of the fingernail bed in that the sensory nerves to the nailbed come from the plantar (palmar) aspect of the foot (hand). Thus, in operating upon an ingrown toenail, the surgeon will insert the needle through the dorsum of the toe base (as the plantar surface is extremely sensitive), but then pass the needle down through the side of the toe to the **plantar** surface, where the anesthesia will affect the plantar nerves. Similarly, in operating upon an infected fingernail bed, it is the palmar side of the finger webbing that must be anesthetized.

In operating upon plantar warts that lie on the sole of the foot, it is painful to inject locally. Instead, a broad area of anesthesia may be obtained by anesthetizing the tibial nerve at the medial malleolus (just superior to the posterior tibial artery pulse).

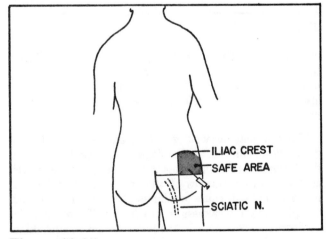

Figure 12-25

CHAPTER 13.   THE AUTONOMIC NERVOUS SYSTEM

The pathways mentioned in the preceeding chapter belong to the **somatic motor** and **somatic sensory** systems. Somatic motor fibers innervate striated skeletal muscle. Somatic sensory fibers innervate predominantly the skin, muscle and tissues other than the viscera. "Viscera" means cardiac muscle, smooth muscle (as in the gut) and glands. Visceral (autonomic) motor nuclei are located between the somatic sensory and motor areas of the spinal cord grey matter (fig. 13-1). It is useful to think of the **autonomic system** as consisting of **visceral sensory** and **visceral motor** components.

The autonomic system regulates glands, smooth muscle and cardiac muscle. It contains sympathetic and parasympathetic components. The **sympathetic system**, as a whole, is a catabolic system, expending energy, as in the "flight or fight" response to danger: e.g.,increasing the heart rate and contractility, and shunting blood to the skeletal muscles and heart. The **parasympathetic system** is an anabolic system, conserving energy: e.g., slowing the heart rate and promoting the digestion and absorption of food.

The cell bodies of preganglionic sympathetic fibers lie in the spinal cord at spinal cord level T1-L2 (fig. 13-2). Spinal cord "level T1-L2" means here the area of the spinal cord in which spinal nerve roots T1-L2 leave the cord. It does not mean vertebrae T1-L2 which lie somewhat lower.

Cell bodies of the parasympathetic system occupy comparable positions at spinal cord level S2-S4. In addition, cranial nerves 3, 7, 9, and 10 have parasympathetic components (3 - pupil and ciliary body constriction; 7 - tearing and salivation; 9 - salivation; 10 - the vagus and its ramifications). The cranial nerves will be discussed in chapter 14.

Proceeding rostrally from the caudal tip of the spinal cord, one first finds a parasympathetic area (S2-S4), followed by a sympathetic region (T1-L2), then parasympathetic areas (CNs 10, 9, 7, 3) and then, successively, a sympathetic and a parasympathetic area, a strange alternating sequence. The latter two areas are the posterior and anterior parts of the hypothalamus (fig. 13-2). The **hypothalamus** acts as a master control over the autonomic nervous system. The cerebrum acts to control the somatic motor system.

Fig. 13-1. A. The sympathetic nerve routes. B. The parasympathetic nerve routes.

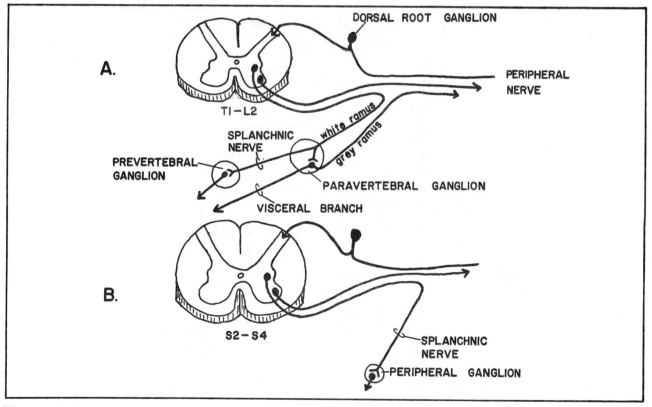

Figure 13-1

Fig. 13-2. Schematic view of the autonomic nervous system.

In figure 13-1 note how certain sympathetic axons follow peripheral nerves, generally to the extremities. It only makes sense to join the peripheral nerve on such a long journey. Why have separate routes when the fibers end up in roughly the same place? E.g., a somatic motor fiber may go to a lumbrical muscle; a somatic sensory fiber may receive sensation from a finger, and a sympathetic motor fiber may innervate a nearby sweat gland or smooth muscle fiber in an arteriole in the finger. Before embarking on such a long trip the sympathetic fibers first stop off in a rest room - synapsing in one of the ganglia of the paravertebral chain. The farther one goes on a trip the dirtier one gets, so the sympathetic axon starts off as a white ramus (myelinated fibers) and continues on as a grey ramus (unmyelinated fibers).

Not all sympathetic axons, however, go to the extremities. Some dive down deeper to the internal viscera in one of two ways:

a. After synapsing in a paravertebral ganglion, they may go directly to the viscera as "visceral" branches. Sympathetic nerves innervating the thorax and pelvis are most prone to do this: e.g., those innervating the heart and lungs, distal colon, and genitalia.

b. Sympathetic fibers innervating the abdominal viscera tend to travel somewhat differently. Rather than synapsing on a **paravertebral ganglion**, they commonly travel right through the ganglion without synapsing. They synapse instead in **prevertebral ganglia** which lie deeper in the abdomen, generally around major blood vessels. Prominent examples of prevertebral ganglia are the **celiac, superior mesenteric** and **inferior mesenteric ganglia**. These are located near the origins of their respective arteries. Nerves which synapse on such deeper-lying ganglia are called **splanchnic nerves**. After synapsing, whether on paravertebral or prevertebral ganglia, sympathetic fibers spread along complex plexuses (usually surrounding major blood vessels) to their target organs.

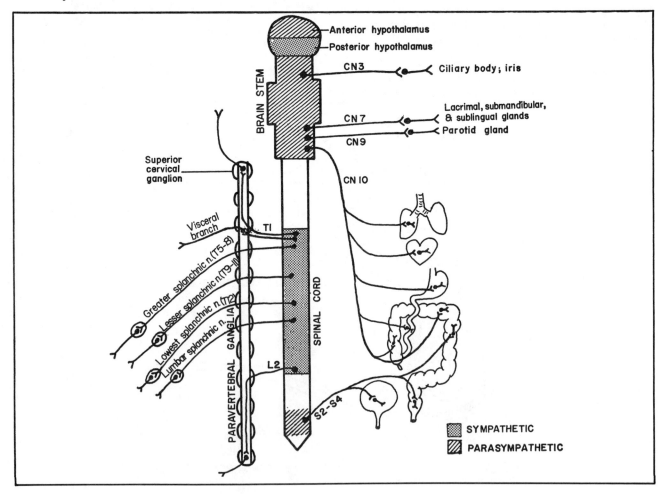

Figure 13-2

The **parasympathetic** system is less complicated. All of its fibers synapse on ganglia that lie near to or within their target organs. Often, they join the same plexuses used by the sympathetic nerves and follow them to their own destinations. Parasympathetic nerves do not supply the extremities. Thus, sympathetic nerves reach virtually the whole body, whereas parasympathetic fibers tend to be restricted to the head, neck and trunk.

In order to extend all over the body, the sympathetic fibers leave the spinal cord at levels T1-L2, enter the paravertebral ganglion chain and then may travel up or down the chain for considerable distances prior to synapsing (fig. 13-2). The sympathetic paravertebral chain lies lateral to the vertebral bodies and stretches from the foramen magnum to the coccyx, supplying the far reaches of the body with sympathetic fibers. Parasympathetic fibers reach widespread areas via the vagus (cranial nerve 10, fig. 13-2).

Most sympathetic fibers to the head and neck travel by way of the **superior cervical ganglion**, the most superior of the paravertebral sympathetic ganglia. This ganglion plus those immediately following (**middle cervical, inferior cervical,** and **thoracic ganglia**) provide important innervation to the cardiac and pulmonary plexuses. The **stellate ganglion** (cervicothoracic ganglion) is actually a fusion of the inferior cervical ganglion and the first thoracic ganglion. It lies on the first and second ribs, posterior to the origin of the vertebral artery.

As a whole, sympathetic fibers to the thorax originate in paravertebral ganglia of the neck and thorax; sympathetic fibers to the abdomen originate in paravertebral ganglia of the thorax and abdomen; sympathetic fibers to the pelvis originate in paravertebral ganglia of the abdomen and pelvis. For example, sympathetic fibers to the celiac, superior mesenteric and inferior mesenteric abdominal (prevertebral) ganglia and smaller related ganglia of the abdomen arise from:

1. The **greater splanchnic nerve** (approx. T5-T8 paravertebral ganglia)
2. The **lesser splanchnic nerve** (about T9-T11 paravertebral ganglion)

3. The **lowest splanchnic nerve** (about T12 paravertebral ganglion)

These preganglionic splanchnic nerves to the abdomen thus arise in the thorax but enter the abdomen by piercing the diaphragm near the aorta. Abdominal prevertebral ganglia also contain the synapses of lumbar sympathetic splanchnic fibers that pass through lumbar paravertebral ganglia without synapsing in the latter.

Both the parasympathetic and sympathetic systems contain two neurons between the spinal cord and periphery. The first synapse is cholinergic (containing acetylcholine). For the sympathetic system this synapse is either in the paravertebral chain of sympathetic ganglia or farther away in prevertebral ganglion plexuses.

The final synapse of the parasympathetic system contains acetylcholine, whereas the final synapse of the sympathetic system contains noradrenaline, with the exception of certain synapses, as for sweating, that contain acetylcholine (i.e., are cholinergic). In figure 13-3, note that secretory functions in general are stimulated by cholinergic fibers.

Fig. 13-3. Autonomic function.

In extreme fear both systems may act simultaneously, producing involuntary emptying of the bladder and rectum (parasympathetic) along with a generalized sympathetic response. In more pleasant circumstances, namely in sexual arousal, the parasympathetic system mediates penile and clitoral erection (probably by arterial vasodilation) and the sympathetic controls ejaculation. Remember: "S2, 3, 4 keeps the penis off the floor".

Visceral pain sensation is generally believed to be transmitted along sympathetic fibers. Visceral pain tends to be poorly localized, as opposed to somatic pain, which commonly localizes precisely. Thus, in early appendicitis, the inflammation, when it involves the visceral peritoneum, localizes poorly, around the umbilicus. When the parietal peritoneum overlying the appendix becomes inflamed, the pain is localized more precisely, to the right lower quadrant of the abdomen.

Structure	Sympathetic function	Parasympathetic function
Eye	Dilates pupil (mydriasis)	Contracts pupil (miosis)
	No significant effect on ciliary muscle	Contracts ciliary muscle (accommodation)
Lacrimal gland	No significant effect	Stimulates secretion
Salivary glands	No significant effect	Stimulates secretion
Sweat glands	Stimulates secretion (cholinergic fibers)	No significant effect
Heart		
Rate	Increases	Decreases
Force of ventricular contraction	Increases	
Blood vessels	Dilates cardiac & skeletal muscle vessels	No significant effect
	Constricts skin and digestive system blood vessels	
Lungs	Dilates bronchial tubes	Constricts bronchial tubes
		Stimulates bronchial gland secretion
Gastrointestinal tract	Inhibits motility and secretion	Stimulates motility and secretion
GI sphincters	contracts	relaxes
Adrenal medulla	Stimulates secretion of adrenaline (cholinergic fibers)	No significant effect
Urinary bladder	?	Contracts
Sex organs	Ejaculation	Erection

Figure 13-3

CHAPTER 14. THE CRANIAL NERVES

There are 12 cranial nerves. Ribald mnemonics are insufficient for remembering them. They are important enough to know on instant recall:

CN1: Smells
CN2: Sees
CNs 3,4 and 6: Move eyes; CN3 constricts pupils, accommodates (focuses the lens for near vision)
CN5: Chews and feels front of head
CN7: Moves the face, salivates (submandibular and sublingual glands), cries, tastes (anterior 2/3 of the tongue),receives small amount of sensation around the external ear
CN8: Hears, receives vestibular input that helps control balance
CN9: Swallows, salivates (parotid gland), tastes (posterior 1/3 of tongue), monitors carotid body and sinus, receives small amount of sensation around the external ear
CN10: Swallows, lifts palate, talks, communicates to and from thoraco-abdominal viscera, tastes (epiglottis), receives small amount of sensation around the external ear
CN11: Turns head, lifts shoulders
CN12: Moves tongue

These functions are more easily remembered by learning the scheme in figure 14-1.

Sympathetic nerves (from the superior cervical ganglion) are not cranial nerves but may join the cranial nerves for short distances. They travel from the neck to the head along the carotid artery and other blood vessels. They elevate the eyelid slightly through Muller's smooth muscle (fig. 15-1), dilate the pupil, and stimulate sweating.

Lesions of sympathetic nerves (**Horner's syndrome**) thus may result in **ptosis, miosis and anhydrosis.**

Fig. 14-1. General organization of the cranial nerves. This figure is a simplified clinically pertinent schema of the functions of the cranial nerves. Somatic motor means innervation of skeletal muscle. Visceral motor means innervation of smooth muscle, cardiac muscle, or glands. Visceral sensory means sensation **returning** from the viscera; taste and smell are included because these functions are associated with the digestive tract. Somatic sensory implies sensation from the skin, skeletal muscle, and associated connective tissues. Unlike the spinal nerves which are mixed nerves (containing motor and sensory components) the cranial nerves are much simpler: Three of the cranial nerves are purely sensory, 5 are purely motor and only 4 are mixed. Also note the following in figure 14-1:

1. Of the 5 motor nerves, only one, CN3, has a visceral motor component (innervation of pupil and ciliary body muscles).
2. The mixed nerves contain all 4 functional motor and sensory components except for CN5, which HAS NO VISCERAL SENSORY AND NO VISCERAL MOTOR component. I.e., CN5 does NOT mediate taste and does NOT mediate any glandular (lacrimal or salivary) functions.

Fig. 14-2. Positioning of the cranial nerves along the brain stem (basal view of brain). For orientation, note that CNs 3 and 6 lie on either side of the pons. Note also that Willis' body (the basilar artery, figs. 6-21,6-22)

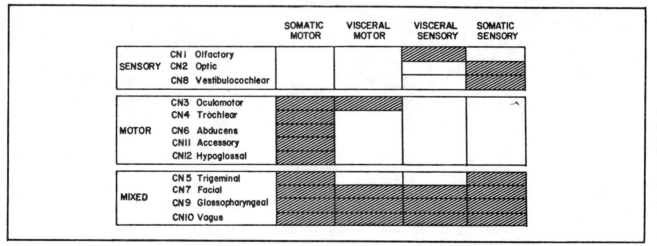

			SOMATIC MOTOR	VISCERAL MOTOR	VISCERAL SENSORY	SOMATIC SENSORY
SENSORY	CN1	Olfactory			▨	
	CN2	Optic				▨
	CN8	Vestibulocochlear				▨
MOTOR	CN3	Oculomotor	▨	▨		
	CN4	Trochlear	▨			
	CN6	Abducens	▨			
	CN11	Accessory	▨			
	CN12	Hypoglossal	▨			
MIXED	CN5	Trigeminal	▨			▨
	CN7	Facial	▨	▨	▨	▨
	CN9	Glossopharyngeal	▨	▨	▨	▨
	CN10	Vagus	▨	▨	▨	▨

Figure 14-1

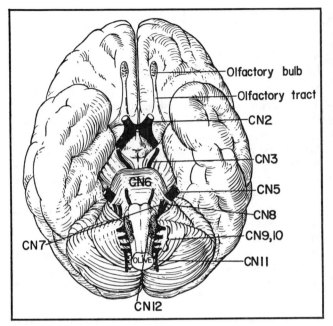

Figure 14-2

stretches from one end of the pons to the other and that CN3 represents Willis' hairy armpits.

Fig. 14-3. The 3 sensory cranial nerves.

Fig. 14-4. The 5 motor cranial nerves (schematic basal view of brain stem).

Fig. 14-5. The 4 mixed cranial nerves. The visceral motor fibers that exit the brain stem are all parasympathetic. Thus the motor ganglia on which they synapse are all located very close to or within their end organs, as is characteristic of the parasympathetic system. The ciliary ganglion, for instance, lies just behind the eye. The peripheral vagal ganglia lie within the end organs.

Just as the posterior root ganglia of the spinal nerves all lie within the protective confines of the vertebral column, all the sensory ganglia of the cranial nerves (whether somatic or visceral sensory) lie within the bony confines of the skull. The only exception is the inferior (visceral) ganglion of the vagus nerve, which lies outside the skull (apparently visceral sensation is considered "inferior" to somatic sensation). It is as if the long vagus nerve was so heavy that it pulled its visceral sensory ganglion outside the skull.

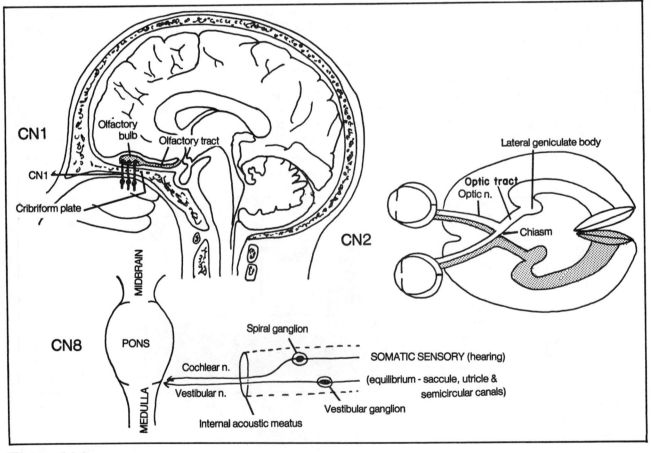

Figure 14-3

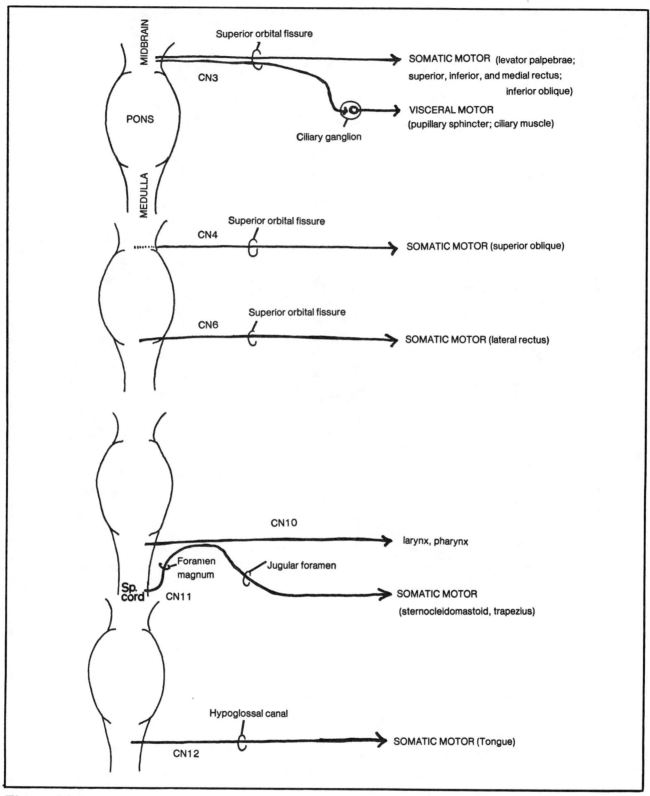

Figure 14-4

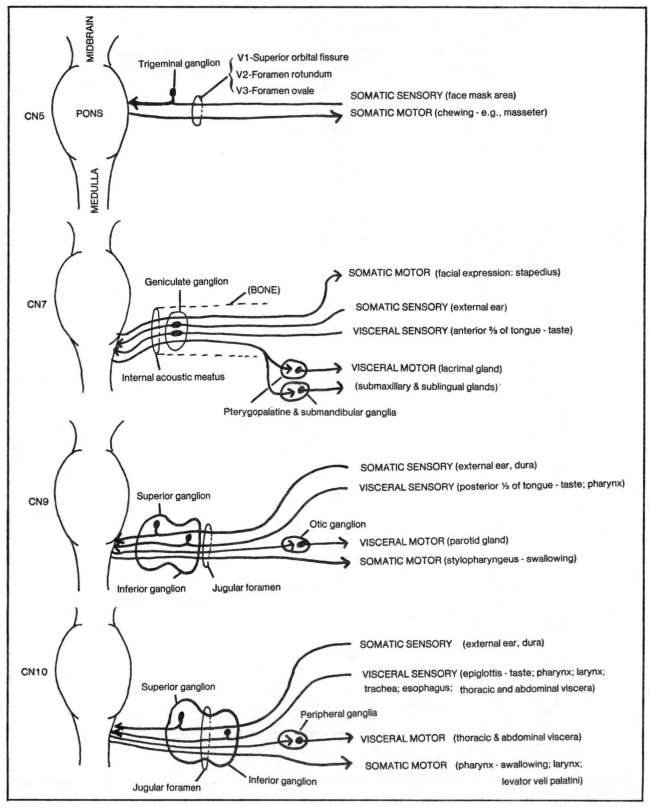

Figure 14-5

CN1, the olfactory nerve, lies nowhere near the brain stem. It extends from the nose through the cribriform plate to connect with the olfactory bulb, where the nerve ends (fig. 14-3). The olfactory nerve impulses are then relayed from the olfactory bulb to the brain via the olfactory tract (fig. 14-3).

CN2, the optic nerve, becomes the optic tract after reaching the brain at the optic chiasm (fig. 14-3).

All the nerves to the eye (CNs, 3,4,5,6, and sympathetics) enter the orbit via the superior orbital fissure, except for the optic nerve which enters the optic canal.

CN4 is not seen in figure 14-2 as it is the only cranial nerve to exit the opposite side, i.e., the roof, of the brainstem (at the midbrain level).
CN5 enters the middle of the pons.

A tumor in the angle between pons, medulla, and cerebellum (cerebellopontine angle tumor) may compress **CNs 7 and 8,** causing deficits in hearing and balance (CN8) and, sometimes, facial paralysis (CN7).

CNs 9, 10 and 12 lie on either side of a bulge in the medulla called the inferior olive (fig. 14-2).

CN11 is not really a cranial nerve. It originates in cervical segments C1-C6, goes up through the foramen magnum and simply touches CN10 (that portion of CN10 that goes to the deep throat - the larynx and pharynx). It then comes right back down again through the jugular foramen to innervate the sternocleidomastoid and trapezius muscles. Some anatomists consider some fibers of CN11 to enter CN10 and facilitate innervation of the throat. This is largely a matter of the semantics of which is a CN10 fiber and which is a CN11 fiber. It is easier to remember CN11 as having the sole function of innervation of the sternocleidomastoid and trapezius muscles. Thus, CN10 is the only nerve to innervate the larynx; CNs 9 and 10 are the only nerves to supply motor fibers to the pharynx (CN9 innervates the stylopharyngeus muscle and CN10 takes care of the other pharyngeal muscles).

Fig. 14-6. Distribution of the 3 sensory branches of CN5 on the face. Note that the angle of the jaw characteristically is not innervated by CN5 but by C2,3. Thus , if loss of facial sensation involves the angle of the jaw, or lower, suspect a cause other than a lesion of CN5.

The 3 main branches of the trigeminal nerve (Ophthalmic - V1; Maxillary - V2; Mandibular - V3) all exit the skull through the sphenoid bone, via the **superior orbital fissure, foramen rotundum** and **foramen ovale** respectively. Remember a round hole lying in the center, surrounded by a fissure on one side and an oval hole on the other.

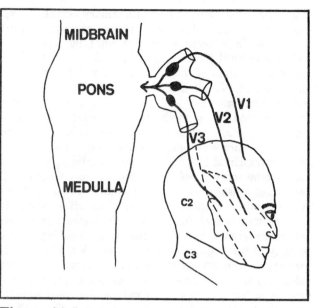

Figure 14-6

Fig. 14-7. Distribution of V1 (ophthalmic nerve). Note how V1 enters the orbit through the superior orbital fissure to innervate the eye (most importantly, the cornea) and orbital contents. It continues on, outside the orbit, to innervate the skin from the scalp to the tip of the nose. Surgeons may anesthetize V1 (**supraorbital branch**) at the supraorbital notch (fig. 2-19), prior to upper lid surgery, as V1 innervates the upper lid. The **lacrimal nerve** supplies sensation to the lacrimal gland and to the lateral upper lid (the skin area overlying the lacrimal gland). The **nasociliary nerve** supplies sensation to the cornea as well as the skin overlying the nose.

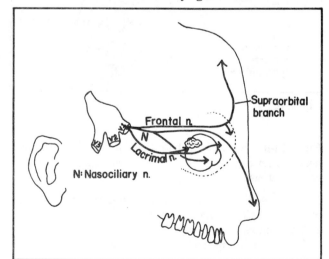

Figure 14-7

Fig. 14-8. Distribution of V2 (maxillary nerve). V2, via the foramen rotundum, enters the **pterygopalatine fossa**, and from there enters the orbit by way of the **inferior orbital fissure** (fig. 2-29). V2 really doesn't do much innervation of the orbit, however. V1 already did that. Instead, V2 runs in the floor of the orbit as the **infraorbital nerve** to exit via the infraorbital foramen and innervate part of the facial skin, including the upper lip. The infraorbital nerve also innervates the lower eyelid, and surgeons sometimes anesthetize it at the level of the infraorbital foramen prior to lower lid surgery. The infraorbital nerve, before leaving the orbit, gives off important sensory branches (**alveolar nerves**) to all the upper teeth and gums. V2 also gives off a relatively unimportant branch to the orbit, the **zygomatic nerve**, which innervates the skin of the temple and face in the vicinity of the zygomatic arch. V1 and V2, within their respective dermatomes, also supply sensory fibers to the nasal passages. Not shown is the **nasopalatine nerve**, a branch of V2 that innervates part of the nasal septum; by extending through the incisive foramen (fig. 3-15), it innervates the anterior aspect of the palate. Dentists may anesthetize the nasopalatine nerve at the level of the incisive foramen.

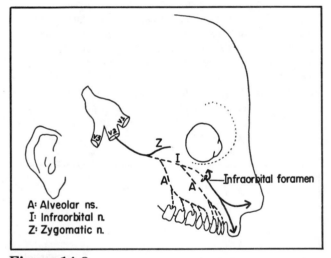

A: Alveolar ns.
I: Infraorbital n.
Z: Zygomatic n.

Figure 14-8

Fig. 14-9. Sensory distribution of V3 (mandibular nerve). **Sensory** fibers of V3 innervate three important general areas:

1. The cheek (both external skin and internal mucous membrane) via the buccal nerve.
2. The tongue, via the lingual nerve
3. The lower gums and teeth, via the inferior alveolar nerve.

The inferior alveolar nerve innervates the teeth, which are imbedded in mandibular bone; naturally there must be an opening in the bone (**mandibular foramen**) with an associated underground tunnel in the bone (**mandibular canal**) to allow access to the teeth. The inferior alveolar nerve eventually exits the mandible via the **mental foramen**, supplying skin of the chin and lower lip. Thus, the chin and lower lip may become numb when the dentist anesthetizes the inferior alveolar nerve. The inferior alveolar nerve runs under the roots of the teeth and can be damaged in extraction of the third molar (wisdom tooth). The **lingual nerve** needs no bony canal, as it innervates the tongue (anterior 2/3 of the tongue) and adjacent floor of the mouth and lower gums. It simply runs under the mucosa of the floor of the mouth. Like the inferior alveolar nerve, it runs close to the third molar and may be damaged during extraction of this tooth. It may also be damaged in tonsillectomy procedures (so can the glossopharyngeal nerve). The lingual nerve typically is anesthetized along with the inferior alveolar nerve during dental procedures, causing numbness of the tongue.

Within the V3 dermatome: the **mental nerve** innervates chin skin, the **buccal nerve** supplies cheek skin, and the **auriculotemporal nerve** innervates the temple.

Motor Fibers Of CN5

All the motor fibers of CN5 lie in V3. They exit the brain stem along with the sensory fibers of CN5 and accompany V3 until V3 is outside the skull, when it is time to get off the road and ride a bucking bronco (fig. 14-10).

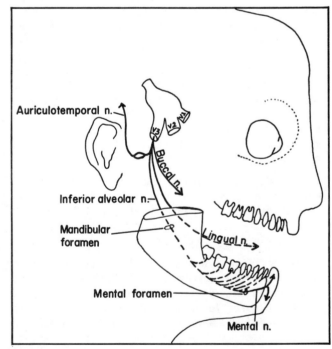

Auriculotemporal n.

Buccal n.

Inferior alveolar n.

Mandibular foramen

Lingual n.

Mental foramen

Mental n.

Figure 14-9

Fig. 14-10. Motor fibers of CN5 presented as a rider on a horse (mandible). The saddle is the **mandibular notch.** The rider's arms and legs are motor nerves that connect with the chewing muscles (**masseter, temporalis, lateral and medial pterygoid**). As this is really a mandible and NOT a BUCking bronco, V3 does NOT innervate the BUCcinator muscle (the buccal branch of CN7 does). The buccal nerve of V3 is a sensory nerve. The rider is TENSE and fibers of V3, in addition to the chewing muscles, innervate the only two TENSor muscles of the head - the **tensor palati** (helps close the nasopharynx in swallowing, and opens the auditory tube) and the **tensor tympani** (helps dampen excessively loud sounds). These nerves are the moustache of the rider and they lie near the otic ganglion (the rider's nose). Other motor fibers of CN5 (not shown in fig. 14-10) continue along the inferior alveolar nerve and at an opportune moment leave it to innervate the **mylohyoid muscle** and the **anterior belly of the digastric muscle.**

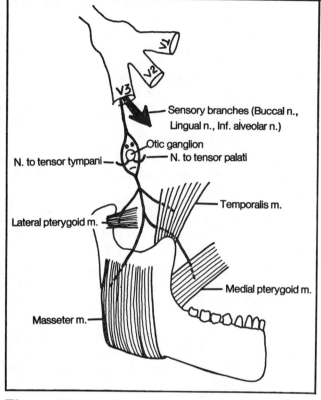

Figure 14-10

The Facial Nerve (CN7)

The facial nerve has 4 types of components (fig. 14-1):

1. Somatic motor - facial musculature
2. Visceral sensory - taste in anterior 2/3 of the tongue
3. Somatic sensory - small amount of skin around the external ear canal
4. Visceral motor - lacrimal and salivary (submandibular and sublingual) glands; mucosal glands of nose and palate.

After it exits the brain stem, CN7 travels a secret route through the temporal bone, from the **internal acoustic meatus** to the **stylomastoid foramen.** All of its branches are given off in this subterranean chamber, except for the somatic motor fibers to the facial muscles, which exit the stylomastoid foramen (figs. 14-11,14-12). Ophthalmologists use this information in anesthetizing the orbicularis muscle (to prevent the patient from squeezing the lids shut and extruding the contents of the eye during intraocular surgery). They pass a needle between the jaw and mastoid bone to anesthetize CN7 where it leaves the stylomastoid foramen.

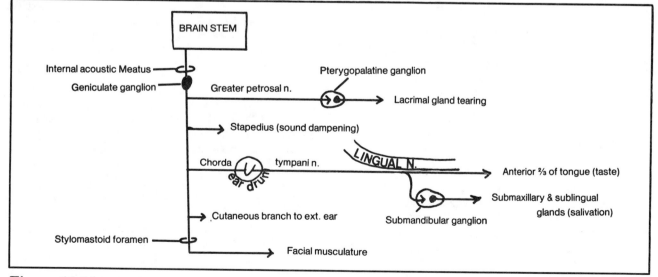

Figure 14-11

Fig. 14-11. Schematized branches of CN7. If a patient presents with facial paralysis and also has other signs of CN7 damage (e.g., decreased taste, excessive loudness of sound from stapedius muscle inaction), probably the injury is proximal to the stylomastoid foramen, where all these respective branches lie. Injury outside the stylomastoid foramen would result only in facial muscle paralysis. The latter paralysis may occur, for example, with a parotid gland tumor (usually a malignant one, as benign ones generally don't compromise the facial nerve).

Fig. 14-12. The major motor branches of the facial nerve on the face. Picture in your mind the key landmarks of the face in profile: the eye, nose, lips, and chin. The key branches lie on either side of these structures. Note that the **buccal branch**, which goes to the buccinator muscle, among other things, is not the same as the buccal nerve (V3 branch of CN5), that goes to the skin and mucous membranes of the cheek.

The cervical branch of CN7 supplies the platysma muscle (fig. 4-57). Sometimes, after neck surgery, there may be difficulty in moving the corner of the mouth on one side. If this is due to severance of the platysma muscle, the condition will likely be temporary, improving with healing of the muscle. It may be permanent, however, if the cervical branch of CN7 is interrupted.

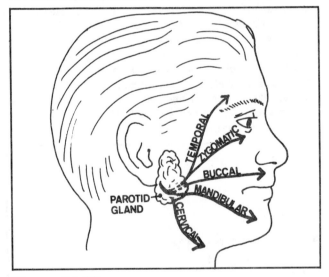

Figure 14-12

Fig. 14-13. The strange fusion of parasympathetic ganglia of CNs 3, 7, and 9 to CN5. Compare with figures 14-7 through 14-9. There are parasympathetic motor ganglia associated with each of the three major branches of CN5 even though CN5 has no autonomic function: V1 - **ciliary ganglion**; V2 - **pterygopalatine ganglion**; V3 - **otic** and **submandibular ganglia**. What are these ganglia doing there if CN5 is not parasympathetic? This

strange situation can be viewed most simply by recognizing that many of the fibers of CNs 3, 7 and 9 are going to places similar to those of the sensory fibers of CN5. The different nerves may, for parts of their courses, take common pathways in order to conserve space, just as two autos going to different destinations may travel along a common road for part of the distance. For instance, consider CN3. Postganglionic parasympathetic fibers of CN3 go to the eye (to the ciliary body and pupillary constrictor, from the ciliary ganglion). Sensory fibers of CN5 also go to the eye, especially to the cornea. CNs 3 and 5 travel close together, near enough for the ciliary ganglion to fuse with CN5. Beyond the ciliary ganglion, fibers from CN3 and CN5 travel together for a short distance via ciliary nerves before separating to go to their final terminations on the eye.

Similarly, postganglionic parasympathetic motor fibers of CN7 go to the lacrimal gland and mucosal glands of the nose and palate (from the pterygopalatine ganglion). Sensory fibers of CN5 also go to the lacrimal gland (via the lacrimal nerve) and mucosa of the nose and palate, so the pterygopalatine ganglion (CN7) fuses to CN5. This allows the postganglionic fibers of CN7 to:

1. briefly hitch **a** ride aboard the zygomatic branch of V2, which connects with the lacrimal nerve of V1, to reach the lacrimal gland.
2. follow CN5 sensory fibers into the nose and palate.

Postganglionic parasympathetic fibers of CN9 supply the parotid gland (from the otic ganglion). It so happens that certain branches of V3 (motor to tensor tympani and tensor palati; the auriculotemporal nerve to the skin of the temple) lie near the parotid gland. Thus, the otic ganglion lies near these branches, and its postganglionic branches accompany the auriculotemporal nerve for a short distance. The otic ganglion itself doesn't actually fuse to the auriculotemporal nerve. It is a little reluctant to do so, because the auriculotemporal nerve does a dangerous stunt. It splits to go around the middle meningeal artery (fig. 14-13).

In summary, CN5 is a large nerve going to many places. The ciliary ganglion (CN3), pterygopalatine ganglion (CN7) and the otic ganglion (CN9) realize this and each associate with a branch of CN5 so that their terminal **postganglionic** branches can steal a ride aboard the terminal parts of V1, V2, and V3.

The situation is slightly different for the submandibular ganglion (CN7): **preganglionic** fibers of CN7 fuse to the trigeminal nerve, as follows. CN7 mediates taste and salivation (submandibular and sublingual glands). Fibers of CN7 wish to go either to the tongue (taste fibers) or near the tongue (salivary fibers to submandibular and sublingual glands). The chorda tympani branch of CN7 travels behind the ear drum and this nerve carries taste and salivatory fibers (figs. 4-63,14-11). The ear drum is a rather strange place for fibers of this type to be. A little voice in the ear whispers to the chorda tympani nerve: "Listen, you're way off your target. You want to get to the tongue region for taste (anterior 2/3 of tongue) and salivation (submandibular and sublingual glands)? I suggest you try the lingual nerve (V3), as it goes to the anterior 2/3 of the tongue for touch sensation." And so, the chorda tympani joins the lingual nerve. The taste fibers follow the V3 touch fibers directly to the tongue, and the salivatory fibers get off near the tongue to supply the submandibular ganglion and its salivary glands. Thus the lingual nerve is a special example, where sensory fibers of CN5 are fused to **preganglionic** parasympathetic fibers (of CN7). The other examples mentioned above are cases where **postganglionic** fibers follow CN5 for short distances. Severing the lingual nerve at any point will result in loss of touch sensation to the tongue. Taste and salivation will be affected, as well, if the distal end is cut beyond the point where the chorda tympani joins the lingual nerve. Cutting the chorda tympani may result in loss of taste but sometimes does not, as taste fibers may also take a detour. They may travel back along the lingual nerve to reach the geniculate ganglion (fig. 14-11) by a devious route

(through the greater petrosal nerve). You can't really blame some of the taste fibers for taking other routes; the chorda tympani runs behind the ear drum and it is embarrassing for a taste fiber to be found running in the ear.

CN10 (The Vagus Nerve)

CN10 is the only nerve innervating the laryngeal muscles. Only two nerves supply the pharyngeal muscles - CN10 and CN9, and CN9 has only a very small function in this area - motor control of the stylopharyngeus muscle. Thus, CN10 injury may significantly impare speech and swallowing. CN10 also innervates all the muscles of the palate (except for the tensor palati; remember, V3 supplies the TENSor muscles, including the tensor tympani and tensor palati). CN10 also relays sensory information back from the viscera (visceral sensory) as well as from a small area of skin in the vicinity of the ear (somatic sensory).

Sometimes, a sore throat may present as a pain in the ear, but no ear pathology is evident. This referred pain occurs because CNs 9 and 10 supply sensation not only to the pharynx but to a portion of the external ear.

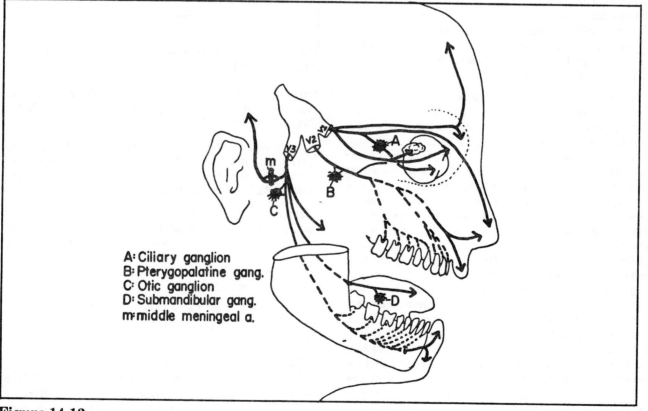

A: Ciliary ganglion
B: Pterygopalatine gang.
C: Otic ganglion
D: Submandibular gang.
m: middle meningeal a.

Figure 14-13

Speech and Swallowing (the somatic motor component of CN10)

Fig. 14-14. Branches of the vagus nerve (CN10).

(A) auricular branch - cutaneous to external ear
(B) pharyngeal branch - to pharyngeal plexus, which supplies all muscles of the pharynx and soft palate except stylopharyngeus (CN9) and **tensor palati** (CN5)
(C) superior laryngeal n. - divides into internal and external branches (letters C1 and C2)

(C1) internal branch of superior laryngeal n. - sensory to mucosa above vocal cords

(C2) external branch of superior laryngeal n. - motor to inferior constrictor m. and cricothyroid m. It travels with the superior thyroid artery for part of its course. Care must be taken not to damage this nerve branch in operations in which the **superior** thyroid artery is tied off. The patient will develop a monotone.

(D) recurrent laryngeal n. - motor to all muscles of the larynx except cricothyroid. It accompanies the inferior larngeal branch of the inferior thyroid artery for part of its course. The recurrent laryngeal nerve may be damaged in operation in which branches of the **inferior** thyroid artery are tied off, causing hoarseness.
(E) cardiac branches
(F) esophageal plexus
(G) anterior vagal trunk to anterior stomach and liver
(H) posterior vagal trunk to posterior stomach and then (via celiac plexus) to liver, kidney, small intestine and large intestine to splenic flexure

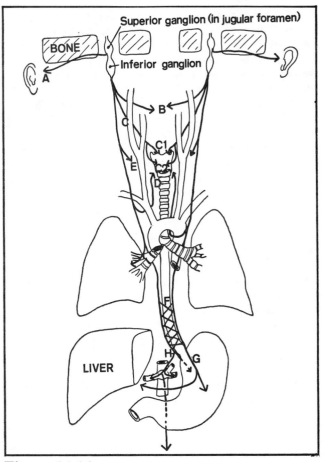

Figure 14-14

There are 3 constrictor muscles of the pharynx - superior, middle and inferior. There are also 3 main vagal branches to the pharynx - the pharyngeal branch, the superior laryngeal nerve and the recurrent laryngeal nerve, all of which contribute to innervation of the pharyngeal muscles. The superior and recurrent laryngeal nerves also innervate the laryngeal muscles. The right recurrent nerve wraps around the front of the subclavian artery and the left recurrent nerve wraps around the aorta (just posterior to the ligamentum arteriosum - an important point for thoracic surgeons operating in this area). The recurrent nerves then backtrack to the larynx as if they forgot to do something very important and went back to complete the task. That task is indeed important. The recurrent nerves innervate all of the muscles of the larynx, except for **cricothyroid**, which is innervated by the **superior laryngeal nerve**. It avoids this muscle because the cricothyroid muscle is the only laryngeal muscle external to the larynx, and the recurrent nerve runs internal to the larynx. It is the **external branch** of the superior laryngeal nerve that innervates the cricothyroid muscle. The **internal branch** of the superior laryngeal nerve goes internal to the larynx, but having no muscle left to innervate, it does supply sensation to the laryngeal mucosa ABOVE the level of the vocal cord.

Since the recurrent laryngeal nerve runs internal to the larynx, it not only innervates many muscles but simultaneously provides sensation to the laryngeal mucosa BELOW the vocal cord.

Injury to the recurrent laryngeal nerve may occur during thyroid or carotid artery surgery, resulting in hoarseness. It may also be compromised by tumors of the neck and lung apex. Bilateral lesions may cause interference with respiration if the vocal cords are closed together at the midline.

The thyroid gland has a fibrous capsule. Surgeons like to stay within the capsule as the recurrent laryngeal nerve lies outside.

The vagus nerve also gives off branches to the heart, lungs, and gastrointestinal tract (see fig. 13-3 for details of function). During their course along the esophagus, the two vagus nerves form many anastomoses with each other. On entering the abdomen through the esophageal opening of the diaphragm, the two mixed nerves are called the **anterior and posterior vagal trunks**, derived mainly from left and right **vagal nerves** respectively. The rotation of the stomach causes the vagal trunks to enter as they do - anteriorly and posteriorly. Surgeons may selectively cut branches of the vagus during ulcer surgery to attempt to decrease gastric acid production. The anterior vagal trunk sends branches to the anterior stomach and the liver. The posterior vagal trunk supplies the posterior stomach; it also anastomoses extensively with the celiac plexus. Thus, the celiac plexus contains a mixture of sympathetic and parasympathetic fibers. Beyond the celiac plexus the vagal fibers distribute to the abdominal viscera, up to the splenic flexure of the colon.

The **hypoglossal nerve** (CN12) innervates the intrinsic and extrinsic muscles of the tongue, except for palatoglossus which is innervated by CN10. (Remember, CN10 innervates all the palatal muscles except for tensor palati which is innervated by CN5). Those who are familiar with the American system of rating movies may find it strange that an X-rated nerve (CNX) innervates a PG muscle (palatoglossus).

Since the cervical nerves arise immediately after CN12, it is not unreasonable for CN12 and C1 to follow one another for some distance. They do so (fig. 12-9) above the level of the hyoid bone. Note that CN12 carries no taste fibers.

CHAPTER 15. THE EYE

It is difficult to play billiards using eyeballs, as the eye is not perfectly spherical. The cornea is too steeply curved (fig. 15-1). This steep curvature enables the cornea to perform most of the **refraction** (bending and focusing) of light entering the eye. The cornea provides a coarse, nonvariable focus. The lens also focuses light, but only performs the fine variable adjustments. Contact lenses artificially alter the curvature of the front of the eye, thereby changing the focus.

Fig. 15-1. Sagittal section through the eye. Arrows indicate the flow of aqueous humor from the ciliary body to the posterior chamber (P), to the anterior chamber (A.C.), to the angle (A), through the filtering (trabecular) meshwork (dotted lines) to the canal of Schlemm(S). A man in the eye is looking at the lens (see fig. 15-2).

(A) angle of anterior chamber
(A.C.) anterior chamber

(C) conjunctiva
(E) eyelash
(F) fornix
(I.R.) inferior rectus m.
(L) levator palpebrae superioris m. (innervated by CN3)
(M) Meibomian gland
(Mu) Muller's m. (innervated by sympathetic nerve fibers from the superior cervical ganglion)
(O) orbicularis m.
(P) posterior chamber
(S) Schlemm's canal
(S.R.) superior rectus m.
(T) tarsal plate
(X) a bad area of the lens to get a cataract

Fig. 15-2. Rear view of the man in figure 15-1. Note that the muscles of the ciliary body form a ring.

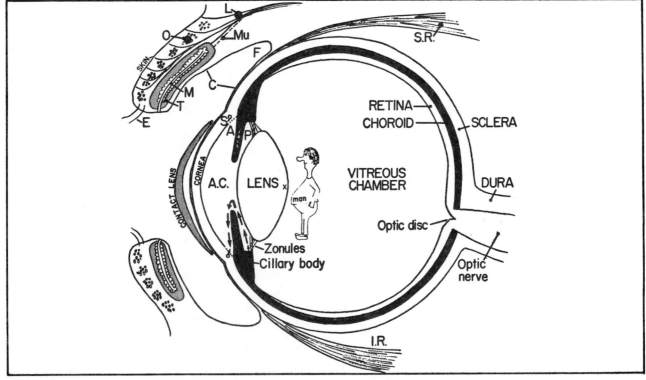

Figure 15-1

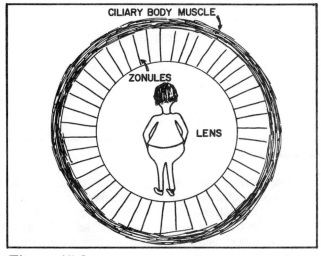

Figure 15-2

Fig. 15-3. Layers of the cornea.

Although only about 1mm thick, the cornea is tough. Following ocular trauma, its outer layer, the corneal **epithelium** (fig. 15-3), regenerates rapidly with little scarring. Most trauma does not penetrate the tough **Bowman's membrane** that lies below the epithelium. Unfortunately, trauma that does affect the corneal **stroma** results in scarring. Particularly impaired vision results when such scarring involves the center of the cornea (in line with the pupil). Corneal transparency is based primarily on the geometric array of collagen fibers in its stroma. Damage to the corneal **endothelium**, which affects the water balance in this collagen meshwork, will result in corneal clouding.

The eye has three chambers: the **anterior chamber** (in front of the iris), the **posterior chamber** (between the iris and the lens), and the **vitreous chamber** (behind the lens). The anterior and posterior chambers contain the clear, watery **aqueous humor**, produced constantly by the **ciliary body**. Aqueous humor exits the eye via the circular **canal of Schlemm**, which lies in the angle between the cornea and iris. The canal of Schlemm communicates directly with the venous system. Blockage of aqueous outflow results in increased intraocular pressure, termed **glaucoma**.

The ciliary body not only produces aqueous humor, but also contains a ring of ciliary muscles (fig. 15-2) that connect to the lens via fine, ligamentous zonule fibers. Contraction of the ciliary muscles affects the shape of the lens, thereby changing its focus - the process of **accommodation**.

The mechanism of accommodation is more easily understood by first explaining the mechanism of

pupillary expansion (dilation) and constriction, as follows. The iris contains circular (constrictor) muscles at the pupillary border, and radial (dilator) muscle fibers (fig. 15-4).

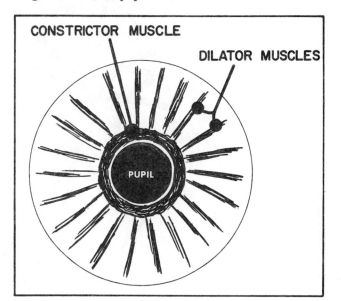

Figure 15-3

Fig. 15-4. Muscles of the iris. Constrictor muscles are innervated by **parasympathetic** fibers of cranial nerve 3 (the oculomotor nerve). Dilator muscles are innervated by **sympathetic** fibers from the superior cervical ganglion of the neck. Note in figure 15-4 how contraction of the radial fibers would dilate the pupil. Contraction of the constrictor muscles decreases the circumference of the ring; therefore, the pupil constricts.

Figure 15-4

The ciliary muscles of the lens, although more complex than the iris constrictor muscles, in a sense act similarly. Contraction of muscles in the ciliary ring narrows the diameter of the ring. This decreases the tension of the zonules, and releases tension on the lens. The lens then thickens (fig. 15-5), leading to a stronger focus (accommodation). Thus, accommodation is the process in which the ciliary muscles contract, thereby relaxing tension on the lens and enabling one to focus closer on an object.

Fig. 15-5. Constriction of the ring of ciliary muscles narrows the diameter of the ring. This reduces tension on the lens. The lens becomes more convex and focuses the light closer to the lens (accommodation).

Opacities in the lens (**cataracts**) may obstruct vision, particularly when positioned centrally, in the posterior aspect of the lens (fig. 15-1). For optical reasons, cataracts in the anterior aspect of the lens or at the lens periphery tend to cause less visual loss.

The vitreous chamber contains **vitreous humor, a thick gel.** Unlike the aqueous humor, vitreous humor is no longer produced in the mature eye. Vitreous that is lost inadvertently from the eye during intraocular surgery cannot be replaced; fortunately, normal saline or aqueous humor may be substituted.

The eye has three main coats - the **retina, choroid** (a very vascular, pigmented structure), and **sclera** (the avascular "white of the eye"). After passing through the

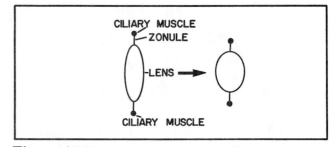

Figure 15-5

cornea, anterior chamber, pupil, posterior chamber, lens and vitreous chamber, light then strikes the transparent retina. The retina contains photoreceptor cells that convert light energy to neuronal impulses which spread along a chain of neurons to the optic nerve, which extends to the brain.

Fig. 15-6. Schematic view of the retina as seen directly through the pupil. The optic cup is the depressed central region of the optic disc where blood vessels and nerve fibers leave the eye to enter the optic nerve. The optic nerve angles in toward the midline optic chiasm. Therefore, the optic disc, as might be expected, lies slightly nasally in the retina, fitting the alignment of the optic nerves. The fovea lies roughly in the center of the retina, about two disc diameters temporal to the optic disc. Four general groupings (arcades) of blood vessels spread to four quadrants of the retina. Veins are slightly

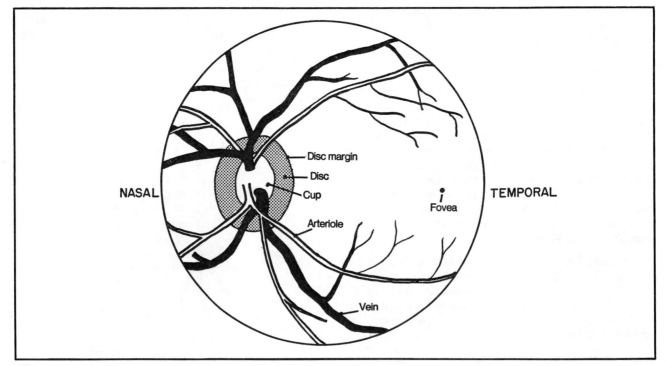

Figure 15-6

larger than arterioles. Nerve fibers from all regions of the retina extend toward the optic disc (optic nerve head), but do not cross the fovea, the area of retina with the most acute vision (fig. 15-7). Normally, the optic axons are transparent, and viewing the retina through an ophthalmoscope will reveal only the structures shown in figure 15-6.

Fig. 15-7. The patterns of optic nerve fibers converging upon the optic disc. The dark sector is the area of decreased optic fiber functioning that might result from a small lesion at the optic disc. Such defects are common in glaucoma.

"Uvea" refers to the combination of the **choroid, ciliary body and iris.** All these structures are pigmented and continuous with one another. If one were to remove the sclera, one would note with great shock that the entire underlying eye, not just the iris, is pigmented. **Uveitis** is an inflammation of the uvea. It is "posterior" when the choroid is involved, or "anterior" when the ciliary body

or iris is involved. **Chorioretinitis** is an inflammation that affects both choroid and retina.

Fig. 15-8. The lacrimal apparatus. Tears travel from the lacrimal gland to the nose.

For muscles of the eye see figures 4-60 through 4-62.

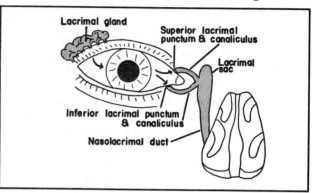

Figure 15-8

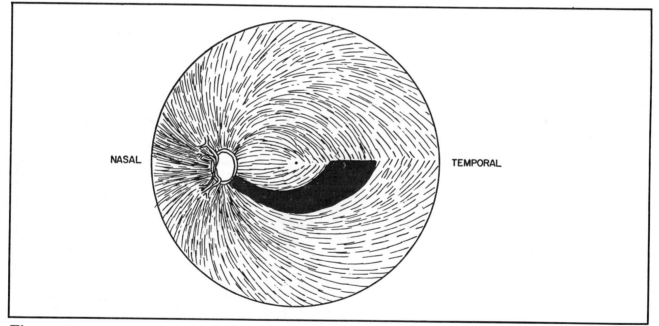

NASAL TEMPORAL

Figure 15-7

CHAPTER 16. THE EAR

The organ of hearing and the organ of vestibular sense (which helps control balance) are intimately associated and will be considered together.

Fig. 16-1. Coronal view of the auditory system. The outer ear canal (external acoustic meatus) extends anteromedially. This slight anterior directionality is important to remember in angling the earpieces of one's stethoscope and in positioning the otoscope for an ear examination.

An explorer crawling into the external acoustic meatus would reach a dead end at the ear drum (tympanic membrane), which marks the transition from outer to middle ear. The eardrum is tilted as if it would fall on top of the explorer. The drum is cone-shaped, seemingly prevented from falling by its attachment to the chain of 3 small bones (ossicles) of the middle ear cavity.

These bones (review fig. 4-63) are:

1. The malleus (hammer), which attaches directly to the drum by its handle.
2. The incus (anvil)
3. The stapes (stirrup).

Fig. 16-2. Explorer's view of the eardrum as seen from the external ear canal. The explorer would like to enter the middle ear. The safest place to cut through the eardrum is inferoposteriorly. There are relatively few blood vessels there, it is out of the way of the ossicles, and

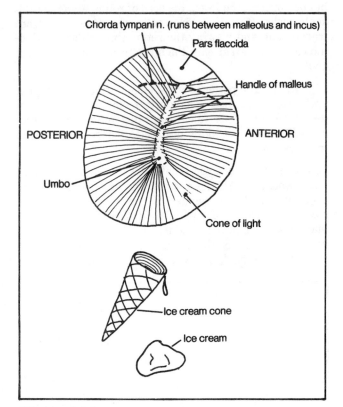

Figure 16-2

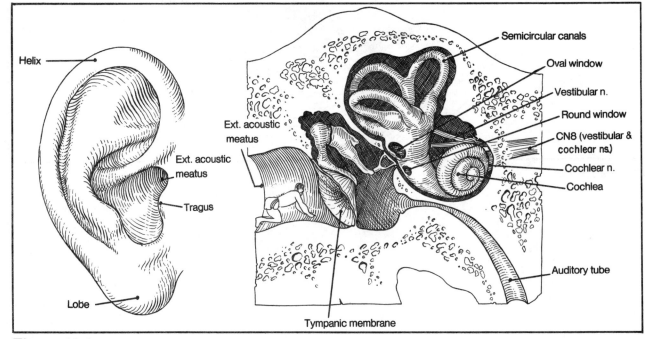

Figure 16-1

the **chorda tympani** nerve (which mediates taste and salivation - see also fig. 4-63) is out of the way. The eardrum is commonly pierced in this inferoposterior region to allow drainage of fluid that has collected in the middle ear. The **cone of light** is a light reflection seen on the anterior aspect of the lower drum. It is important to note this reflection on exam, as it may disappear when fluid collects behind the ear drum. For orientation, one way to remember the location of the cone of light and the direction of tilt of the handle of the malleus is to imagine an ice cream cone. The ice cream cone is tilted anteriorly toward the mouth, causing the ice cream to fall to the ground (too bad; now the chorda tympani, which lies superiorly, can't taste it).

Once in the middle ear, the explorer would see overhead the three ear bones, vibrating rapidly and relaying sound impulses from the tympanic membrane to the **oval window**. The oval window is a hole in the bone that marks the transition from middle ear to inner ear. The stapes attaches to the oval window, behaving like a nervous door knocker that vibrates very rapidly. Actually, the 3 bones look like a muscle-man standing on top of the tympanic membrane and about to swat the explorer with a tennis racquet (fig. 16-1). Fortunately, the tennis racquet is stuck to the oval window.

The explorer could leave the middle ear by sliding down the **auditory (Eustachian) tube** in the anteroinferior aspect of the middle ear (fig. 16-3) or by climbing up into a hole in the roof, the **mastoid antrum.** The explorer opts, instead, to go scuba diving. Rather than leave through air-filled passages, he dons a scuba outfit, cuts out the stapes, enters the oval window, and swims into the inner ear, specifically into the **perilymph** of the **vestibule** (fig. 16-3). The scala VESTIBULI gets its name from the fact that it connects directly with the vestibule, or entrance, to the inner ear. The PERIlymph gets its name from the fact that it occupies the outer, or more peripheral of the two fluid filled chambers (the other containing endolymph).

Fig. 16-3. The inner ear. Normally, the inner ear (cochlea) is rolled up like a snail shell (fig. 16-1). It is here schematically unrolled as if someone blew out a New Year's Eve noisemaker.

Sound waves passing across the oval window travel in the scala vestibuli and are transmitted to the **cochlear duct** where they stimulate nerve fibers that connect with the **organ of Corti.** Impulses from the organ of Corti (see cross section in fig. 16-4) travel along **cochlear nerve** fibers, in company with vestibular nerve fibers from the saccule, utricle and semicircular canals, to the brain stem.

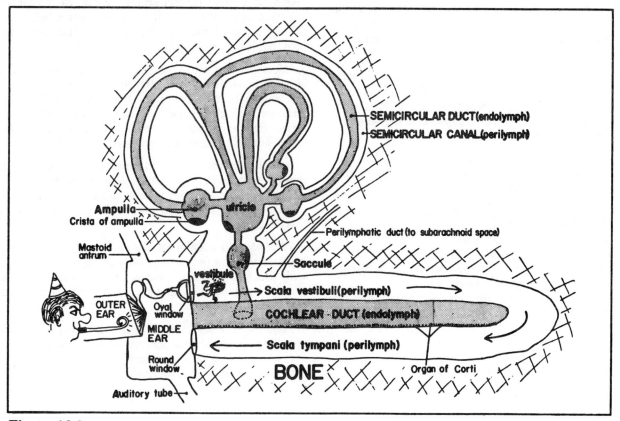

Figure 16-3

The three **semicircular canals** are shaped somewhat like a pretzel. These three canals lie at right angles to one another. Their three ampullae contain neurosensitive structures (**cristae**) that are sensitive to movements of the head. Note that the semicircular canals connect with the utricle in five places.

The **saccule** and **utricle** each contain a **macula**, a motion-sensitive structure which helps determine the position of the head and provides information about acceleration and deceleration.

If the diver continues swimming in the perilymph, he will reach the **scala tympani** (a continuation of the perilymph that will carry the diver full circle back toward the tympanic cavity (middle ear). The diver will not enter the tympanic cavity because the **round window** obstructs the way. The round window has no ossicles attached to it and thus is not distorted out of shape like the oval window. Piercing through the round window, the explorer would reenter the middle ear. If the explorer wished, he could have entered the subarachnoid space of the brain via the **perilymphatic duct** or swum around in the perilymph of the utricle, saccule, and semicircular canals. He would not, however, have swum in any **endolymph** which occupies a separate internal chamber (like a balloon within a balloon). The endolymph, like the perilymph, may be found in the cochlea (specifically in the cochlear duct) as well as in the saccule, utricle, and semicircular ducts. The endolymph fluid has a different chemical composition than the perilymph.

The explorer was wise in leaving through the round window, rather than exploring the vestibular system for that would take him through a SEWER system (**SUA=Saccule, Utricle, and Ampulla of the semicircular canals**).

Fig. 16-4. Cross section through: A. A semicircular canal, B. The cochlea. Both have a "balloon-within-a-balloon" pattern. Compare with figure 16-3.

(1) semicircular canal (contains perilymph as do the scala vestibuli and scala tympani)
(2) semicircular duct (endolymph)
(3) vestibular membrane (angled, like the side of a "V")
(4) tectorial membrane (moves and stimulates neurosensitive hair cells of the organ of Corti)
(5) scala Tympani (its top runs straight across, like the top of a "T")
(6) organ of Corti
(7) spiral ganglion
(8) cochlear nerve
(9) vascular stria (produces endolymph)

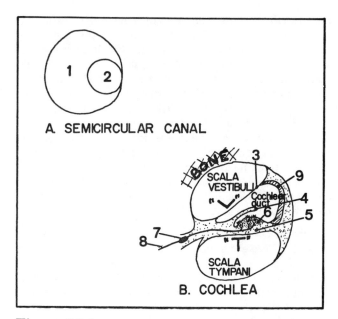

Figure 16-4

If the explorer were to place his ear against the floor of the middle ear, he would hear a thumping, as the pulsating internal carotid artery (as well as the jugular vein) lies below the floor. If he were to cut a hole in the ceiling he would enter the attic (intracranial cavity) and note that it was insulated by dura.

Infections may enter the middle ear (**otitis media**, as opposed to **otitis externa**, which is an inflammation of the external auditory canal) by spreading from the pharynx through the auditory tube. Patients with a sore throat may experience ear pain. Sometimes this is due to spreading infection and sometimes to referred pain.

Infection in the middle ear may spread through the mastoid antrum to the mastoid sinuses (**mastoiditis**). The closeness of the sigmoid venous sinus (fig. 6-41) to the middle ear may allow entry of middle ear infections into this sinus.

Patients with **Meniere's disease** experience recurrent bouts of **vertigo** (a spinning sensation), often associated with **tinnitus** (ringing, buzzing or other sounds), which are believed to be related to excessive endolymph pressure.

Patients with **otosclerosis**, a not infrequent concomitant of aging, suffer hearing loss from loss of mobility of the otic ossicles. Surgery, such as implantation of an artificial stapes, is often successful in improving hearing.

CHAPTER 17. REGIONAL POINTS

Triangles Of the Neck

Fig. 17-1. The anterior and posterior triangles of the neck. The letter "N" defines the anterior and posterior triangles, as viewed from the right. The bases of the triangles consist of bones: the clavicle for the posterior triangle, and the mandible for the anterior triangle.

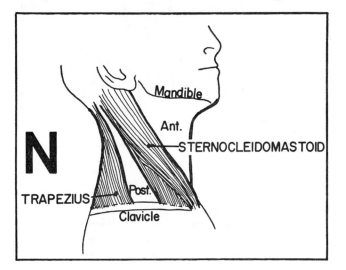

Figure 17-1

Fig. 17-2. Subdivisions of the anterior and posterior triangles. Subdivided triangles are: Occ., occipital; Subcl., subclavian; Car., carotid; Mus., muscular; Subma., submandibular; Subme, submental. Imagine someone on a bus pulling the cord by the window to signal the bus driver to stop. The arm and forearm are the posterior and anterior bellies of the omohyoid muscle (the elbow is hidden behind the sternocleidomastoid). The hand (hyoid bone) is grabbing the digastric muscle at the junction between its posterior and anterior bellies. This subdivides the triangles into 6 more triangles, each with special qualities, as follows:

Divisions Of the Posterior Triangle (fig. 17-2)

1. The occipital triangle - contains very important nerves particularly in its lower aspect. CN11 (accessory nerve), which supplies the sternocleidomastoid and trapezius muscles, crosses the occipital triangle between these muscles (fig. 12-10). A danger zone lies inferior to CN11: the brachial plexus and phrenic nerve lie in this region, extending between scalenus anterior and medius (figs. 12-14 and 17-3). The major sensory nerves to skin of the neck come out of hiding behind the sternocleidomastoid muscle to supply the neck (fig. 12-10). A stab wound in the lower aspect of the occipital triangle could have serious neurologic consequences. There might be a fair amount of bleeding too, from interruption of the external jugular vein (fig. 6-39).

2. The subclavian triangle - contains the subclavian artery (fig. 12-14). Compression of this zone may prove useful in controlling upper extremity bleeding.

Fig. 17-3. Deep muscles of the posterior triangle. A number of "S" muscles occupy the posterior triangle, in alphabetical order! The muscles between Sternocleidomastoid and Trapezius (i.e. between "S" and "T") are, alphabetically:

SCAlenius Anterior
SCAlenius Medius
SCAlenius Posterior
levator SCAPulae
SPlenius capitis

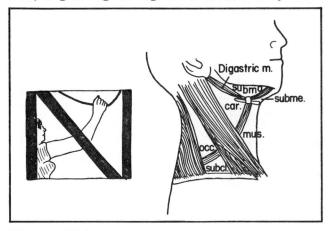

Figure 17-2

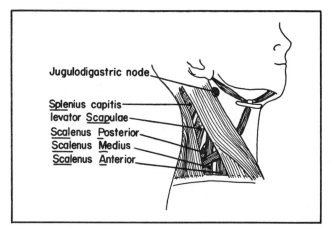

Figure 17-3

The above muscles are shown individually in figures 4-24 (sternocleidomastoid), 4-2 (trapezius, levator scapulae), 4-21 (scalene muscles), and 4-17 (splenius).

Divisions Of the Anterior Triangle (fig. 17-2)

1. Carotid triangle - contains the carotid sheath with enclosed carotid artery, internal jugular vein, and vagus nerve. The **tonsilar (jugulodigastric)** node lies at about the angle of the jaw, below the level of the posterior belly of the digastric muscle (fig. 17-3). It commonly enlarges in tonsilitis. The common carotid pulse is frequently palpated in the carotid triangle, to check circulation to the brain.

2. Submental triangle - the only unpaired triangle (fig. 17-2,4). In order to prevent the contents of the mouth, such as the tongue, from falling out between the mandible and hyoid bone, the mylohyoid muscle (fig. 4-81) forms a diaphragm that largely fills this triangle. Submental lymph nodes may become involved with carcinoma that spreads from the anterior aspect of the jaw and lower lip.

3. Submandibular triangle - contains the submandibular gland, hypoglossal nerve (remember that this nerve passes to the tongue superior to the hyoid bone) and submandibular lymph nodes which may enlarge with metastases from carcinoma of the lip.

4. Muscular triangle (fig. 17-2) - contains the strap muscles of the neck, which include two long muscles (omohyoid, sternohyoid), and two short muscles (thyrohyoid, sternothyroid). See also figure 17-4.

Fig. 17-4. Frontal view of muscular triangles of the neck. The heavy lines outline the right and left muscular triangles. The sternohyoid muscle has been remove on the left side of the neck to show the deeper lying thyrohyoid and sternothyroid muscles.

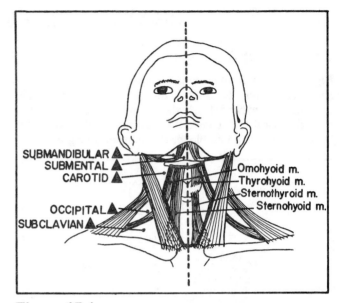

Figure 17-4

Fig. 17-5. The axilla (horizontal section and lateral view). The first rib is the boundary between neck and axilla. The axilla is the space that lies deep to the skin of the armpit. The walls of the axilla are paneled:

1. anteriorly by the pectoralis major (felt in the anterior skin fold).
2. posteriorly by the latissimus dorsi and the teres major muscles (felt in the posterior skin fold), and the scapula.
3. medially, by the serratus anterior muscle, which lies against the rib cage.

Fig. 17-5 A. The quadrangular and triangular spaces. These are three spaces produced by the 3 "T"s: Triceps, Teres major, and Teres minor. Note that the axillary nerve passes through the quadrangular space to wrap

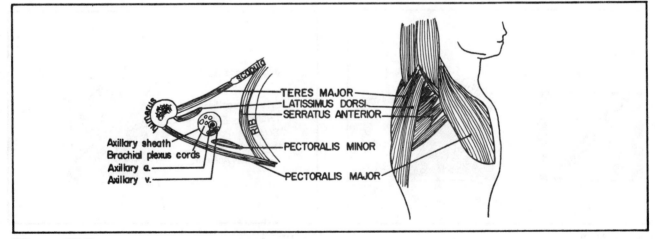

Figure 17-5

around the surgical neck of the humerus. Fractures of the surgical neck may thus injure the axillary nerve.

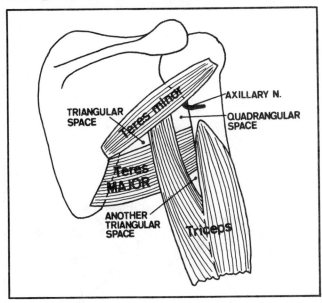

Figure 17-5A

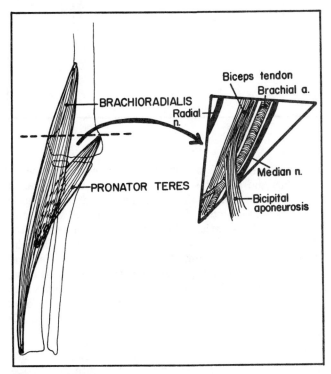

Figure 17-6

Fig. 17-6. The cubital fossa, anterior view. The cubital fossa is a roughly triangular area, bound by the brachioradialis muscle laterally, the pronator teres muscle medially and by an imaginary line between the humeral epicondyles. There is little danger in drawing blood from surface veins that cross this triangle, because the veins lie superficially, away from vital structures. However, in withdrawing arterial blood from the deeper-lying brachial artery (e.g., for blood gas analysis) one would not like to stab away blindly, as the median and radial nerves also lie deeply. The biceps muscle and tendon (which are palpable) lie relatively centrally. The radial nerve lies lateral to the biceps muscle, and the median nerve lies medial. The pulse may be palpated just medial to the biceps muscle and tendon. The bicipital aponeurosis separates the brachial artery from the more superficial veins, a useful point to remember in performing a venous cutdown procedure.

Fig. 17-7. The wrist. The cross section is at the level of the distal radius. A suicide attempt, by wrist slashing, that only extends to the superficial veins is unlikely to cause serious neuromuscular damage. Arterial hemorrhage (i.e. spurting as opposed to oozing) implies deeper damage, and important nerves or tendons may be involved. Cut #1, on the radial side, would damage four main structures:

1. the radial artery (prior to its dive into the anatomical snuff box), or its superficial palmar branch, which joins the superficial palmar arch
2. the flexor carpi radialis (wrist will then deviate to the ulnar side on attempted flexion)
3. the median nerve (causing difficulty with thumb opposition)
4. palmaris longus (causing weakness in wrist flexion)

Cut #2, on the ulnar side, would sever three main structures:

1. the ulnar artery
2. flexor carpi ulnaris (causing deviation of the wrist radially on attempted flexion)
3. the ulnar nerve (causing paralysis of most of the small hand muscles - claw hand).

With deeper midline incisions, the tendons of flexor digitorum superficialis as well as the deeper flexor tendons might be severed, causing varying degrees of difficulty in finger flexion.

Note that the flexor retinaculum attaches to the 4 corners of the two rows of carpal bones.

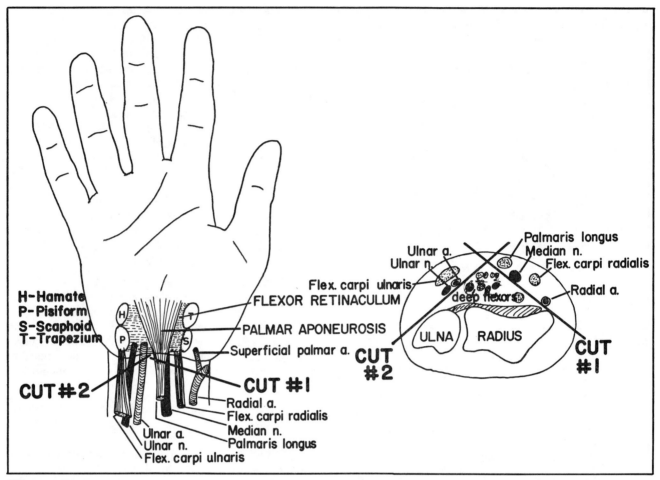

H-Hamate
P-Pisiform
S-Scaphoid
T-Trapezium

FLEXOR RETINACULUM
PALMAR APONEUROSIS
Superficial palmar a.

CUT #2

CUT #1
Radial a.
Flex. carpi radialis
Median n.
Palmaris longus

Ulnar a.
Ulnar n.
Flex. carpi ulnaris

Ulnar a.
Ulnar n.
Flex. carpi ulnaris

deep flexors

ULNA   RADIUS

Palmaris longus
Median n.
Flex. carpi radialis

Radial a.

CUT #1

CUT #2

Figure 17-7

Fig. 17-8. The femoral triangle. This triangle is bound by adductor longus and the long sartorius muscle, and the inguinal ligament. The navel lies at the midline of the abdomen. Similarly, the NAVL (sequence of femoral Nerve, Artery, Vein, and Lymphatics) extends from lateral toward the midline in the femoral triangle. In drawing arterial blood from the femoral artery, palpate the pulse and insert the needle directly. In drawing venous blood from the femoral vein, palpate the arterial pulse and insert the needle medial to the pulse (not laterally, where one may injure the femoral nerve). Fortunately the femoral nerve is somewhat separated from the other structures, lying outside the femoral sheath that binds artery, vein and lymphatics in a separate canal. In femoral hernias, a portion of bowel may herniate under the inguinal ligament into the canal of the femoral sheath.

Figure 17-8 also shows the course of the femoral artery on leaving the femoral triangle. The femoral artery - and

vein (not shown) - travel along, sandwiched between sartorius and adductor longus muscles in what is called the adductor canal. The adductor magnus muscle contains a hole (hiatus) that allows the femoral artery and vein to shift posteriorly behind the knee to enter the popliteal fossa and become the popliteal artery and vein. The femoral nerve initially accompanies the femoral artery but never makes it through to the popliteal fossa, as it is too busy giving off branches in the anterior thigh. Instead, the sciatic nerve occupies the popliteal fossa, reaching the latter by a totally different route (fig. 12-24).

The femoral artery bisects the femoral triangle. It also separates the motor territories of the obturator nerve from the motor territories of the femoral nerve. Atheromatous occlusion of the femoral artery is relatively common at the adductor hiatus.

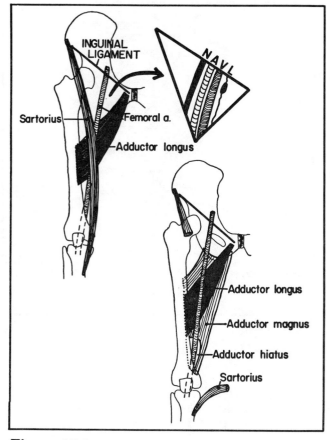

Figure 17-8

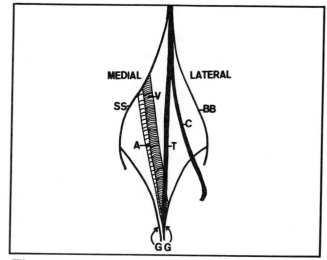

Figure 17-9

Fig. 17-10. Two fierce soccer players. Relations of the major arterial and venous trunks. The defensive Aorta comes from behind and extends his legs in front of the Vena Cava, attempting to abort Vena's offensive attack. Note that above the umbilicus the major arterial branches tend to lie posterior to the major venous branches. The opposite is true below the umbilicus. Note that the right common iliac artery crosses the left common iliac vein. Sometimes this causes venous compression and varicose veins in the left lower extremity.

Fig. 17-9. The right popliteal fossa (posterior view). SS, semitendinosus and semimembranosus mm.; BB, biceps femoris, long and short heads; GG, gastrocnemius, medial and lateral heads. Unlike the triangular cubital fossa, the popliteal fossa is diamond-shaped. It contains the popliteal artery(A), popliteal vein(V), and the main components of the sciatic nerve - the tibial(T) and common peroneal(C) nerves.

In the old days, a warrior would disable the enemy by cutting the hamstrings (tendons of biceps, semitendinosus, and semimembranosus, preventing the ability to run. A cut across the hamstrings might also damage the tibial nerve. A deeper cut might cause significant venous hemorrhage by severing the deeper-lying popliteal vein. A still deeper cut would cause massive hemorrhage as the popliteal artery lies deepest. One may feel the popliteal pulse by palpating deeply.

Not shown in figures 17-8 and 17-9 are the small and large saphenous veins. The small saphenous vein is a laterally located superficial vein that joins the popliteal vein in the popliteal fossa, whereas the great saphenous vein is a medially located superficial vein that joins the femoral vein in the femoral triangle (see fig. 6-43).

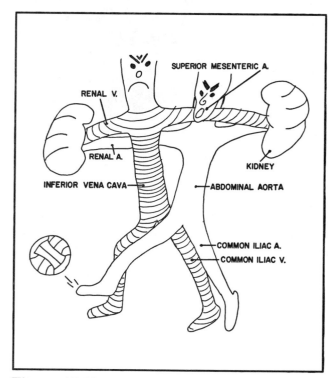

Figure 17-10

Fig. 17-11. Relationships of the trachea and bronchi to the aorta, azygous vein and pulmonary arteries. The pulmonary arteries drape over the bronchi like an exhausted person whose arms and body are being supported by a friend (Bronco Billy). The moon (azygous arch in cross section) and sun (aortic arch in cross section) may be seen over the right and left shoulders, respectively, of Bronco Billy. The azygous arch joins the superior vena cava. The aortic arch originates from the heart and passes over the right pulmonary artery (fig. 6-1) to become the descending aorta.

Fig. 17-12. Relationships of trachea, aorta, and esophagus. Bronco Billy in figure 17-11 has his back up against a post (esophagus) but the esophagus shifts from a posterior to an anterior position as it extends downward to join the stomach.

The tracheal rings are discontinuous posteriorly. Thus, it is relatively easy for esophageal malignancies to erode into the trachea.

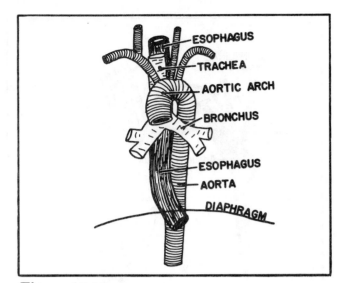

Figure 17-12

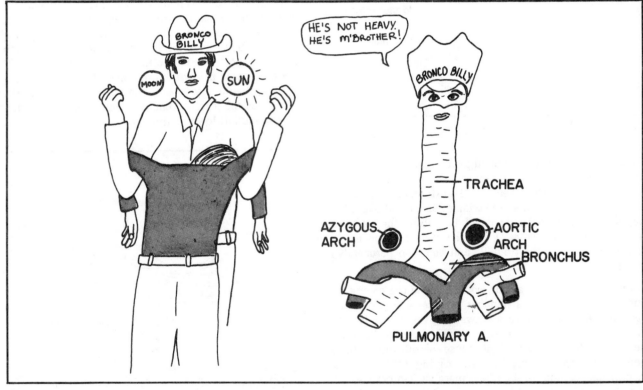

Figure 17-11

170

Fig. 17-13. Relations of the ureter and nearby arteries. The external iliac artery rests on the psoas major muscle and is crossed by the ureter. It is important to avoid damage to the ureter when tying off the ovarian or uterine vessels as the ureter runs close to these blood vessels. At the level of the uterus, the ureter runs just under the uterine artery ("water runs under the bridge").

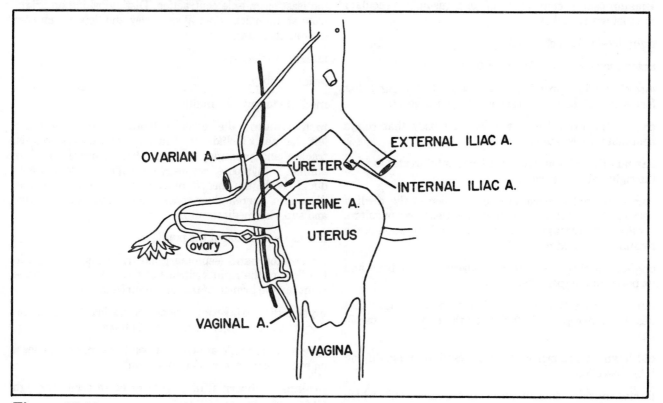

Figure 17-13

GLOSSARY

anastomosis - connection of one (usually vascular) channel with another.

anhydrosis - lack of sweating.

arteriosclerosis - hardening of the arteries.

auricle (atrial appendage) - the pouch ("little ear") that forms part of the wall of each atrium of the heart.

axon - the extension of a nerve cell body that relays neuronal impulses to another cell.

carina - the point at which the trachea bifurcates to form the right and left bronchi.

carpal tunnel syndrome - compression of the median nerve at the wrist, commonly by the flexor retinaculum, resulting in decreased motor function and sensation in the thumb, index, and middle fingers.

cephalhematoma - bleeding between a skull bone and its pericranium (periosteum).

cuneiform cartilage - a small nodule of cartilage that lies in the aryepiglottic fold, near the corniculate cartilage.

cutdown - catheterizing a blood vessel after identifying it by dissection.

dartos muscle - smooth muscle in the scrotal skin which wrinkles with cold and elevates the testes.

diplopia - double vision.

ectopic - displaced.

epigastrium - the area of abdomen inferior to the xiphoid process.

evaginate - protrude outward.

fauces - the threshold between mouth and pharynx.

glottis - the space between the vocal cords.

infarction - tissue death secondary to loss of oxygenation.

inion (external occipital protuberance) - the midline bony prominence on the posterior aspect of the occipital bone; the superior origin of the trapezius muscle.

insertion - the muscle attachment point that moves the most on muscle contraction. This contrasts with the muscle "origin", which is the attachment point that moves the least on muscle contraction. Distinguishing an origin from an insertion is not always easy and depends on how the muscle is used.

lateral - to the side.

lesion - injury.

medial - toward the midline.

mediastinum - the region of thorax between right and left pleurae. It includes the heart, associated great vessels, trachea and proximal bronchi, phrenic nerve and four animals - thymoose, esophagoose, vagoose, and thoracic duck. With the heart representing the middle mediastinum, surrounding zones are the anterior, posterior and superior mediastinum.

miosis - narrowing of the pupil.

moderator band (septomarginal trabecula) - a ridge in the floor of the right ventricle of the heart; it carries the right bundle branch of the atrioventricular bundle.

myocardial infarction - death of cardiac tissue, secondary to loss of oxygenation; a heart attack.

origin - the muscle attachment point that moves the least on muscle contraction. See "insertion".

prolapse - abnormal displacement of an organ through an area of weakened support.

sigmoidoscopy - examination of the sigmoid colon and rectum through a viewing tube.

somatic - pertaining to body structures other than smooth muscle, cardiac muscle and glands.

stroke - a prolonged or permanent loss of function in a brain area, resulting from interruption of the blood supply.

varices - abnormally dilated veins.

vasodilation - expansion of blood vessel diameter, usually referring to the arterial system.

visceral - pertaining to smooth muscle, cardiac muscle, or glands; pertaining to the internal organs of the body.

xiphoid process - the sharp inferior border of the sternum (figs. 2-1 and 4-25). Patients feeling this sometimes are unnecessarily alarmed in thinking it is an abdominal tumor.

APPENDIX

Muscle Innervations

The muscle innervations in this list in many cases are approximate. It is difficult and unnecessary to remember all of them in great detail. General patterns of innervation are discussed in chapters 12-14 (NERVOUS SYSTEM).

ABDUCTOR DIGITI MINIMI
(FOOT), lateral plantar n. (S1, 2)
(HAND), ulnar n. (T1)
ABDUCTOR HALLUCIS, medial plantar n. (L5, S1)
ABDUCTOR POLLICIS BREVIS, median n. (T1)
ABDUCTOR POLLICIS LONGUS, radial n. (C7)
ADDUCTOR BREVIS, obturator n. (L2, 3)
ADDUCTOR HALLUCIS, lateral plantar n. (S1, 2)
ADDUCTOR LONGUS, obturator n. (L2, 3)
ADDUCTOR MAGNUS, obturator and sciatic ns. (L3, 4)
ADDUCTOR POLLICIS, ulnar n. (T1)
ANAL SPHINCTER, EXTERNAL, pudendal n. (perineal and inferior rectal branches S3, 4)
INTERNAL, pelvic splanchnic (parasympathetic) ns. (S3, 4)
ANCONEUS, radial n. (C7, 8)
ARYTENOIDS, recurrent laryngeal n.(CN7)
AURICULARES, CN7
BICEPS BRACHII, musculocutaneous n. (C5, 6)
BICEPS FEMORIS, sciatic n. (S1, 2)
BRACHIALIS, musculocutaneous n. (C5, 6)
BRACHIORADIALIS, radial n. (C6)
BUCCINATOR, CN7
BULBOSPONGIOSUS, pudendal n. (S2, 3, 4 perineal branches)
COCCYGEUS, pudendal plexus
COMPRESSOR NARIS, CN7
CONSTRICTOR MUSCLES OF PHARYNX, CN7
CORACOBRACHIALIS, musculocutaneous n. (C7)
CORRUGATOR, CN7
CREMASTER, genitofemoral n. (L1, 2)
CRICOARYTENOIDS, recurrent laryngeal n.(CN10)
CRICOTHYROID, superior laryngeal n., external branch (CN10)
DELTOID, axillary n. (C5)
DEPRESSOR ANGULI ORIS, CN7
DEPRESSOR LABII INFERIORIS, CN7
DEPRESSOR SEPTI, CN7
DIAPHRAGM, phrenic nerve (C3, 4, 5)
DIGASTRIC, ANTERIOR BELLY, CN5
POSTERIOR BELLY, CN7
DILATOR NARIS, CN7
EXTENSOR CARPI RADIALIS BREVIS, radial n. (C6, 7)
EXTENSOR CARPI RADIALIS LONGUS, radial n. (C6, 7)

EXTENSOR CARPI ULNARIS, radial n. (C7, 8)
EXTENSOR DIGITI MINIMI, radial n. (C7)
EXTENSOR DIGITORUM, radial n. (C7, 8)
EXTENSOR DIGITORUM BREVIS, deep peroneal n. (L5, S1)
EXTENSOR DIGITORUM LONGUS, deep peroneal n. (L5, S1)
EXTENSOR HALLUCIS LONGUS, deep peroneal n. (L5, S1)
EXTENSOR INDICIS, radial n. (C7)
EXTENSOR POLLICIS BREVIS, radial n. (C7)
EXTENSOR POLLICIS LONGUS, radial n. (C7)
EXTERNAL OBLIQUE (ABDOMEN), ant. rami T7-T12, L1, 2
ERECTOR SPINAE, posterior rami
FLEXOR ACCESSORIUS, lateral plantar n. (S1, 2)
FLEXOR CARPI ULNARIS, ulnar n. (C7, 8)
FLEXOR DIGIT MINIMI BREVIS
(FOOT), lateral plantar n. (S1, 2)
(HAND), ulnar n. (T1)
FLEXOR DIGITORUM BREVIS, medial plantar n. (L5, S1)
FLEXOR DIGITORUM LONGUS, tibial n. (S1, 2)
FLEXOR DIGITORUM PROFUNDUS, ulnar and median ns. (C8, T1)
FLEXOR DIGITORUM SUPERFICIALIS, median n. (C7, 8, T1)
FLEXOR HALLUCIS BREVIS, medial plantar n. (L5, S1)
FLEXOR HALLUCIS LONGUS, tibial n., (S1, 2)
FLEXOR POLLICIS BREVIS, median and ulnar ns. (T1)
FLEXOR POLLICIS LONGUS, median n. (C8, T1)
GASTROCNEMIUS, tibial n. (S1, 2)
GEMELLUS INFERIOR, ns. to quadratus femoris and gemellus inferior (L5, S1, 2)
GEMELLUS SUPERIOR, ns. to obturator internus and gemellus superior (L5, S1, 2)
GENIOGLOSSUS, CN12
GENIOHYOID, C1 (through CN12)
GLUTEUS MAXIMUS, inferior gluteal n. (L5, S1, 2)
GLUTEUS MEDIUS, superior gluteal n. (L4, 5)
GLUTEUS MINIMUS, superior gluteal n. (L4, 5)
GRACILIS, obturator n. (L2, 3)
HYOGLOSSUS, CN7
ILIACUS, femoral n. (L2, 3)
ILIOCOSTALIS, posterior rami
INFERIOR OBLIQUE (EYE), CN3
INFERIOR RECTUS (EYE), CN3
INFRASPINATUS, suprascapular n. (C5, 6)
INTERCOSTALS, intercostal ns.

INTERNAL OBLIQUE (ABDOMEN), ant. rami T7-T12, L1, 2
INTEROSSEI
   (FOOT), lateral plantar n. (S1, 2)
   (HAND), ulnar n. (T1)
INTERSPINALES, posterior rami
INTERTRANSVERSARII, ant. and post. rami
INTRINSIC TONGUE MUSCLES, CN12
ISCHIOCAVERNOSUS, pudendal n. (S, 2, 3, 4 perineal branches)
LATERAL RECTUS (EYE), CN6
LATISSIMUS DORSI, thoracodorsal n. (C7, 8)
LEVATOR ANGULI ORIS, CN7
LEVATOR ANI, pudendal plexus (S3, 4)
LEVATOR PALPEBRAE SUPERIORIS, CN3
LEVATOR SCAPULAE, C3, 4
LEVATOR SUPERIORIS ALEQUAE NASI, CN7
LEVATOR VELI PALATINI (LEVATOR PALATI), CN10
LEVATORES COSTARUM, anterior rami
LONGISSIMUS, posterior rami
LONGUS CAPITIS, C1-C4 anterior rami
LONGUS COLLI, C2-C8 anterior rami
LUMBRICALES
   (FOOT), L5, S1, 2 through medial and lateral plantar ns.
   (HAND), T1 through median and ulnar ns.
MASSETER, CN5
MEDIAL RECTUS (EYE), CN3
MENTALIS, CN7
MULTIFIDUS, posterior rami
MUSCULUS UVULAE, CN10
MYLOHYOID, CN5
OBLIQUUS CAPITIS INFERIOR, C1,2 posterior rami
   SUPERIOR, C1 posterior rami
OBTURATOR EXTERNUS, obturator n. (L3, 4)
OBTURATOR INTERNUS, ns. to obturator internus and gemellus superior (S1, 2)
OCCIPITOFRONTALIS, CN7
OMOHYOID, ansa cervicalis (C1, 2, 3)
OPPONENS DIGITI MINIMI, ulnar n. (T1)
OPPONENS POLLICIS, median n. (T1)
ORBICULARIS OCULI, CN7
ORBICULARIS ORIS, CN7
PALATOGLOSSUS, CN10
PALATOPHARYNGEUS, CN10
PALMARIS BREVIS, ulnar n. (T1)
PALMARIS LONGUS, median n. (C8)
PECTINEUS, femoral n. (L2, 3)
PECTORALIS MAJOR, medial and lateral pectoral ns. (C5, 6, 7, 8, T1)
PECTORALIS MINOR, medial pectoral n. (C6, 7, 8)
PERONEUS BREVIS, superior peroneal n. (L5, S1)
PERONEUS LONGUS, superior peroneal n. (L5, S1)
PERONEUS TERTIUS, deep peroneal n. (L5, S1)

PIRIFORMIS, S1, 2
PLANTARIS, tibial n. (L5)
PLATYSMA, CN7
POPLITEUS, tibial n. L5
PROCERUS, CN7
PRONATOR QUADRATUS, median n. (C8, T1)
PRONATOR TERES, median n. (C6)
PSOAS MAJOR, L2, 3
PSOAS MINOR, L1, 2
PTERYGOIDS, CN5
QUADRATUS FEMORIS, n. to quadratus femoris (L5, S1)
RECTUS ABDOMINIS, anterior rami T7-T12, L1
RECTUS CAPITIS ANTERIOR, anterior rami C1, 2
   LATERALIS, anterior rami C1, 2
   POSTERIOR MAJOR, posterior ramus C1
   POSTERIOR MINOR, posterior ramus C1
RECTUS FEMORIS, femoral n. (L3, 4)
RHOMBOID MAJOR, dorsal scapular n. (C5)
RHOMBOID MINOR, dorsal scapular n. (C5)
RISORIUS, CN7
ROTATORES, posterior rami
SACROSPINALIS, posterior rami
SALPINGOPHARYNGEUS, CN10
SARTORIUS, femoral n. (L2, 3)
SCALENUS ANTERIOR, anterior rami C5-C8
   MEDIUS, anterior rami C3, 4
   POSTERIOR, anterior rami C3-C8
SEMIMEMBRANOSUS, sciatic n. (L5, S1, 2)
SEMISPINALIS, posterior rami
SEMITENDINOSUS, sciatic n. (L5, S1, 2)
SERRATUS ANTERIOR, long thoracic n. (C5, 6, 7)
SERRATUS POSTERIOR INFERIOR, anterior rami T10-T12
   SUPERIOR, anterior rami T1-T3
SOLEUS, tibial n. (L5, S1, 2)
SPHINCTER URETHRAE, pudendal n. (perineal branches)
SPINALIS, posterior rami
SPLENIUS CAPITIS AND CERVICIS, posterior rami C5-C8
STAPEDIUS, CN7
STERNOCLEIDOMASTOID, CN11 (and C2)
STERNOHYOID, ansa cervicalis (C1-C3)
STERNOTHYROID, ansa cervicalis (C1-C3)
STYLOGLOSSUS, CN12
STYLOHYOID, CN7
STYLOPHARYNGEUS, CN9
SUBCLAVIUS, n. to subclavius (C5, 6)
SUBSCAPULARIS, upper and lower subscapular ns. (C5, 6)
SUPERIOR OBLIQUE (EYE), CN4
SUPERIOR RECTUS (EYE), CN3
SUPINATOR, radial n. (C5, 6)
SUPRASPINATUS, suprascapular n. (C5, 6)

TEMPORALIS, CN5
TENSOR FASCIA LATAE, superior gluteal n. (L4, 5)
TENSOR TYMPANI CN5
TENSOR VELI PALATINI (TENSOR PALATI), CN5
TERES MAJOR, lower subscapular n. (C5, 6)
TERES MINOR, axillary n. (C7, 8)
THYROARYTENOID, CN10 (recurrent laryngeal n.)
THYROEPIGLOTTICUS, CN10 (recurrent laryngeal n.)
THYROHYOID, C1
TIBIALIS ANTERIOR, deep peroneal n. (L4, 5)
TIBIALIS POSTERIOR, tibial n. (L5, S1)
TRANSVERSE PERINEI MUSCLES, pudendal n. (perineal branches)

TRANSVERSUS ABDOMINIS, anterior rami T7-T12, L1
TRANSVERSUS THORACIS, intercostal ns.
TRAPEZIUS, CN11; C3, 4
TRICEPS BRACHII, radial n. (C6, 7, 8)
VASTUS INTERMEDIUS, femoral n. (L3, 4)
VASTUS LATERALIS, femoral n. (L3, 4)
VASTUS MEDIALIS, femoral n. (L3, 4)
VOCALIS, CN10 (recurrent laryngeal n.)
ZYGOMATICUS, CN7

INDEX

NOTE: The words in this index are assigned according to figure number rather than page number. For example, "Brachial vein, 6-39" means that the term "brachial vein" is found in illustration 6-39 and/or in the text between sections 6-39 and 6-40.